The Flowchart Approach to Industrial Cluster Policy

Other books by IDE-JETRO

INDUSTRIAL CLUSTERS IN ASIA
Akifumi Kuchiki and Masatsugu Tsuji (*editors*)

SPATIAL STRUCTURE AND REGIONAL DEVELOPMENT IN CHINA
Nobuhiro Okamoto and Takeo Ihara (*editors*)

GENDER AND DEVELOPMENT
The Japanese Experience in Comparative Perspective
Mayumi Murayama (*editor*)

EAST ASIA'S DE FACTO ECONOMIC INTEGRATION
Daisuke Hiratsuka (*editor*)

RECOVERING FINANCIAL SYSTEMS
China and Asian Transition Economies
Mariko Watanabe (*editor*)

DEVELOPMENT OF ENVIRONMENTAL POLICY IN JAPAN AND ASIAN COUNTRIES
Tadayoshi Terao and Kenji Otsuka (*editors*)

REGIONAL INTEGRATION IN EAST ASIA
From the Viewpoint of Spatial Economics
Masahisa Fujita (*editor*)

NEW DEVELOPMENTS OF THE EXCHANGE RATE REGIMES IN DEVELOPING COUNTRIES
Hisayuki Mitsuo (*editor*)

ECONOMIC INTEGRATION IN ASIA AND INDIA
Masahisa Fujita (*editor*)

The Flowchart Approach to Industrial Cluster Policy

Edited by

Akifumi Kuchiki

and

Masatsugu Tsuji

palgrave macmillan

IDE-JETRO

First published in 2008 by
PALGRAVE MACMILLAN
Houndmills, Basingstoke, Hampshire RG21 6XS and
175 Fifth Avenue, New York, N.Y. 10010
Companies and representatives throughout the world.

PALGRAVE MACMILLAN is the global academic imprint of the Palgrave Macmillan division of St. Martin's Press, LLC and of Palgrave Macmillan Ltd. Macmillan® is a registered trademark in the United States, United Kingdom and other countries. Palgrave is a registered trademark in the European Union and other countries.

ISBN-13: 978–0–230–55361–3 hardback
ISBN-10: 0–230–55361–3 hardback

This book is printed on paper suitable for recycling and made from fully managed and sustained forest sources. Logging, pulping and manufacturing processes are expected to conform to the environmental regulations of the country of origin.

A catalogue record for this book is available from the British Library.

A catalog record for this book is available from the Library of Congress.

2007051222

10 9 8 7 6 5 4 3 2
17 16 15 14 13 12 11 10 09 08

Printed and bound in Great Britain by
CPI Antony Rowe, Chippenham and Eastbourne

Contents

List of Illustrations

Figures

Map

List of Tables

Preface

This volume is the product of the research project entitled "Flowchart Approach to Industrial Cluster Policy" which was coordinated and financially supported by the Institute of Developing Economies (IDE), Japan External Trade Organization (JETRO).

Policy for forming an industrial cluster, or industrial cluster policy, plays an important role in developing a region in East Asia. Industrial cluster policy is followed not only in Japan by the Ministry of International Trade and Industry or the Ministry of Economy, Trade and Industry, but also in most countries in East Asia such as Malaysia and Singapore. Silicon Valley in the United States and Bangalore in India are well-known success stories of development in information technology. Theories of agglomeration and cluster explain that industrial clusters are effective ways in generating external economies and reducing transportation costs in regions as well as whole economies. But it is necessary to clarify what conditions are required for built industrial clusters particularly in developing countries as strategies for economic growth.

The purpose of this book is to present a framework to analyze industrial cluster policies, which is based on the Flowchart Approach initiated by our previous work, *Industrial Clusters in Asia: Analyses of their Competition and Cooperation* [edited A. Kuchiki and M. Tsuji, Palgrave Macmillan, December 2005]. The Flowchart Approach makes it clear what factors are crucial in building industrial clusters, describes how it agglomerates firms, and prioritizes policy measures.

I am grateful to all the members and counterparts who contributed to the project. I also thank all the individuals and organizations that accepted our visits and shared with us their precious time and information.

Any opinions expressed in this book are contributors' but not their organizations'. We hope this publication will contribute to academia and toward further understanding of the Flowchart Approach as a practical method.

AKIFUMI KUCHIKI
Executive Vice President
Japan Trade External Organization

List of Contributors

Akifumi Kuchiki	Executive Vice President, Japan External Trade Organization (JETRO).
Aya Okada	Professor, Graduate School of International Development, Nagoya University.
Hiroyuki Tsukada	Director General, Guangzhou, China, Japan External Trade Organization (JETRO).
Masahisa Fujita	Professor, Konan University and Professor Emeritus, Kyoto University.
Masatsugu Tsuji	Professor, Graduate School of Applied Informatics, University of Hyogo, and Professor Emeritus, Osaka University.
Shoichi Miyahara	Associate Professor, College of Economics, Aoyama Gakuin University.
Yasushi Ueki	Research Fellow, Development Studies Center, Institute of Developing Economies, JETRO (IDE/JETRO).
Yoshiaki Hisamatsu	Associate Professor, Faculty of Regional Development Studies, Toyo University.
Jobaid Kabir	Manager, Environmental Compliance, Lower Colorado River Authority.
Kentaro Yoshida	JETRO, and Visiting Research Fellow, Center for Strategic & International Studies (Washington, DC).
N.S. Siddharthan	Honorary Professor, Madras School of Economics.

Part I

Theoretical Framework

1
Introduction

Akifumi Kuchiki and Masatsugu Tsuji

1. Background

Globalization is not a new phenomenon, but its importance is still growing at an ever faster pace. All economies must either cope with this reality or gradually lose their vitality. If economies successfully and positively respond to this world trend, their development will be ensured.

The economic development of East Asia can be considered a successful example of adaptation to globalization, although the region faced one serious economic crisis in 1997. Most East Asian economies recovered from that crisis, however, and have been rushing forward on a new growth path. A natural question is how these economies can regain their position as the center of world economic growth. Fueled by globalization, many regions in East Asia continue to attract all kinds of economic resources from all over the world, including foreign direct investment (FDI), funding, human resources, and advanced technology and know-how.

As part of what the United Nations Development Project has dubbed the "East Asian Miracle," harmonious relationships between the private and public sectors were emphasized and cultivated. This approach characterizes the famous Japanese industrial policy, in which government guides the private sector by targeting long-term and strategic objectives, seeking to concentrate resources and energy on these. The Korean and Taiwanese electronics industries represent other successful examples of public-private cooperation.

However, the agglomeration that has been occurring in East Asia since the 1990s cannot be attributed to traditional industrial policy. We wish to examine a new type of policy, which is more relevant to East Asia's current phenomena, known as industrial clustering policy. This book represents an attempt to formulate this type of policy.

Here the terms of agglomeration and cluster have the same meaning (the strict definitions will be provided in Chapter 1). The term cluster was utilized extensively by Porter (1998), who proposed applying a "diamond approach" to determine whether a cluster can maintain competitive advantage relative to other clusters (see Introduction of Chapter 3 in more detail). Once a cluster is formed, Porter deems it to be "innovative" if four conditions – factor conditions, demand conditions, related industries, and firm strategy/rivalry – are satisfied. However, this approach does not address the question of how a government can intentionally form a cluster through industrial cluster policy. Baldwin (2003) discussed an industrial cluster policy theory, but did not demonstrate a practical application of the theory. A remaining question is whether an industrial cluster policy can be effective in practice. Answering this question is another aim of this book.

Industrial cluster policy is subtle and complex, requiring not only a traditional combination of targets and policy measures but also related arrangements such as economic reforms, deregulation, construction of infrastructure, establishment of legal systems, and so on. In addition, any successful economy must now meet global standards; without carrying out the reforms necessary for meeting these standards, no region can attract global resources.

Even for developed economies, clustering policy is necessary; Japan is a perfect example. The Japanese economy is only now emerging from a long recession that began in the early 1990s. Since then, countless measures to revitalize the industrial sector have been implemented by all levels of government. A significant amount of public funding has been poured into various projects, including promotion of venture businesses and support for universities through the TLO (Technology Licensing Office), which aims to further university-industry collaboration. However, these policy measures were largely unsuccessful in revitalizing Japanese industry. The lessons learned from these experiences has led to the recognition of the need to harness the power of industrial agglomeration; Japan and other countries must understand that new industries and new businesses do not emerge alone, in isolation from the regional economy, even in developed economies. With the advent of the field of spatial economics, the phenomenon of industrial agglomeration, or clustering, has been the focus of much recent research.

Taking these domestic as well as global circumstances into consideration, our research team focused on the questions of why firms agglomerate in particular regions, and what industrial cluster policies, if any, can be effective in forming industrial clusters. As a group, we have long been pursuing the answer to these important questions, in papers such as Tsuji, Giovannetti, and Kagami (2007) and in Kuchiki and Tsuji (2005). One result of our efforts has been the development of the "flowchart approach," first proposed by Kuchiki (2005). The "flowchart approach" is a synthetic concept that includes theoretical and policy frameworks, and attempts to answer the above two questions. The details of the approach will be explained in the next section.

Thus, the objectives of the research projects detailed in this book are to hypothesize and elaborate the flowchart approach by examining the current situation of clustering in various economies and isolating the common factors that contributed to cluster formation in all of them. Moreover, we attempt to verify empirically the extent to which the flowchart approach can explain clustering phenomena, and to examine whether actual policies aimed at promoting agglomeration in certain regions or in certain industries are effective. In other words, we aim to generalize the flowchart approach so that it can be applicable to any economy or industry. It would not be an exaggeration to state that whether or not a region can survive in the competitive world may depend on the success or failure of industrial cluster policy. However, it is not yet known whether industrial cluster policy can by itself form an effective industrial cluster from scratch.

2. A prototype model of the flowchart approach

Figure 1.1, which appeared first in Kuchiki (2005), clearly indicates the components of the flowchart approach. The flowchart approach posits that industrial cluster policy can be effective in forming industrial clusters by establishing export-processing zones or industrial zones, building capacity, and inviting anchor firms. An anchor firm is defined as a firm that manufactures its products by assembling intermediate goods of parts and

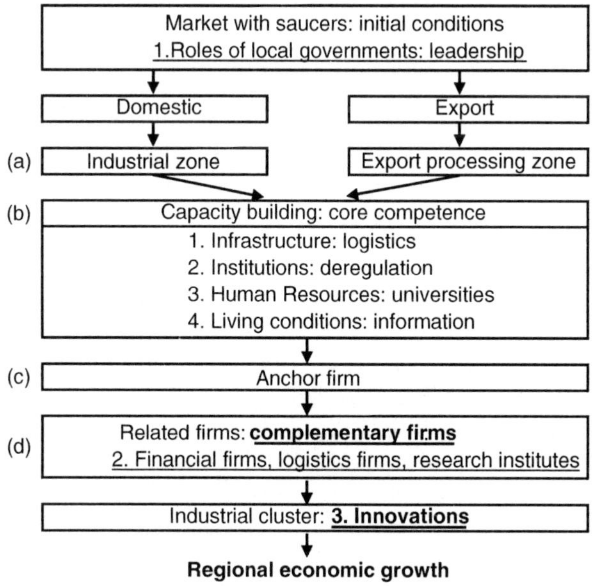

Figure 1.1 Flowchart approach to industrial cluster policy
Source: author.

components. Its relative firms are suppliers in the part and component industries. Capacity building is here defined as facilitation of infrastructure, institutional building, human resources development, and arrangements of living conditions. Appendix of this book proposed sufficient conditions for the formation of the type of industrial clusters common in the Asian manufacturing industry, whose presence enhances regional economic growth.

Typical industrial cluster policy theory defines an industrial zone as a "quasi-public good"; Appendix of this book showed that policies enhance economic growth if an anchor firm operates under a production function of "increasing returns to scale" (see the definition of industrial zone in Chapter 3). Critical levels of production necessary to create "scale economies," used by related firms to decide whether or not to invest in clusters, were also shown. Kuchiki (2007), and Tsuji and Quan (2005) illustrated two cases of successful cluster policy: Canon in Hanoi, Vietnam, and Toyota in Tianjin, China. Canon and Toyota, two Japanese companies, functioned well in those clusters as the "anchor firms" required by the flowchart approach.

The purpose of this section is to apply a prototype model of the flowchart approach to industrial cluster policy in the manufacturing industry. The flowchart approach offers two basic guidelines for the implementation of cluster policy. First, the timing and ordering of policy measures is vital. Second, one must specify the economic agents responsible for building the various types of capacity necessary for industrial cluster policy, choosing from among the actors of central government, local government, various actors in the quasi-public sector, and private firms.

An anchor firm in the manufacturing industry assembles products composed of various parts and components. An assembler in the automobile industry typically uses over 20,000 parts per car, and an anchor firm in the printer industry uses about 800 parts per printer. The anchor firm's related firms generally move into an industrial zone where their anchor firm is a tenant of that zone. If sufficient conditions are met, an industrial cluster will be formed around the industrial zone through agglomeration of the related firms. The agglomeration will bring growth to the region, including the cluster itself.

This pattern has been confirmed in several industrial clusters, including the automobile clusters in Tianjin and Guangzhou in China, the electronics cluster in North Vietnam, the automobile cluster near Bangkok, Thailand, and the electronics clusters in Penang and Johor, Malaysia. It is clear that the presence of an anchor firm and its related firms creates an environment that encourages regional growth.

Let us clarify the scope and the hypothesis of this book.

2.1. Scope

1. This book focuses on anchor firms, or core firms who procure parts and components from suppliers. These firms seek to minimize their total costs, including transportation costs.

2. We apply flowchart models to various manufacturing industries in Asia. Our prototype model differs from industry to industry, and from region to region. The flowchart for the information industry will be different from that of the manufacturing industry.
3. The role of central governments and local governments is capacity building, which prepares an investment environment that is inviting to foreign firms.

2.2. Hypothesis

> A region creates industrial agglomeration by following the approach detailed in the flowcharts of Figures 1.1 and 1.2.

The flowchart approach to industrial cluster policy, shown in Figure 1.1, considers industrial cluster policy to be not a national industrial policy but a regional growth strategy. A sufficient condition for the successful implementation of industrial cluster policy is the satisfaction of the conditions of industrial zones, capacity building, and anchor firms, in the correct order.

Industrial clusters in East Asia typically satisfy the above conditions. A local government first forms an industrial zone as a "saucer" to invite foreign investors. Next, the government builds capacity for those investors. The capacity building includes constructing and facilitating physical infrastructure, developing institutions, developing human resources, and creating living conditions satisfactory to foreign investors. Physical infrastructure includes roads, ports, communications, and others. Institution building is crucial to successfully inviting foreign investors into a region, and may include streamlining investment procedures by providing one-stop services, deregulation, and introduction of preferential tax systems. Human resources include unskilled labor, skilled labor, managers, researchers, and professionals. Living environment includes, for example, the provision of hospitals and schools that meet the needs of foreign workers. An anchor firm will typically be motivated to invest in a region after the above-mentioned capacity is built.

We explain Step I our flowchart approach in Figure 1.2. First, we ask whether industrial zones have been established. If the answer is "No," then we must identify the actors responsible for establishing industrial zones. Generally, the central government establishes industrial zones during a country's early stage of industrialization. As a country's economy matures, other actors may establish industrial zones to attract more foreign investment. For example, in Thailand and Malaysia, actors in the government-affiliated semi-public sector, including local authorities, established export processing zones and free trade zones. In addition, Japanese trading corporations have helped to establish many industrial zones in the ASEAN countries.

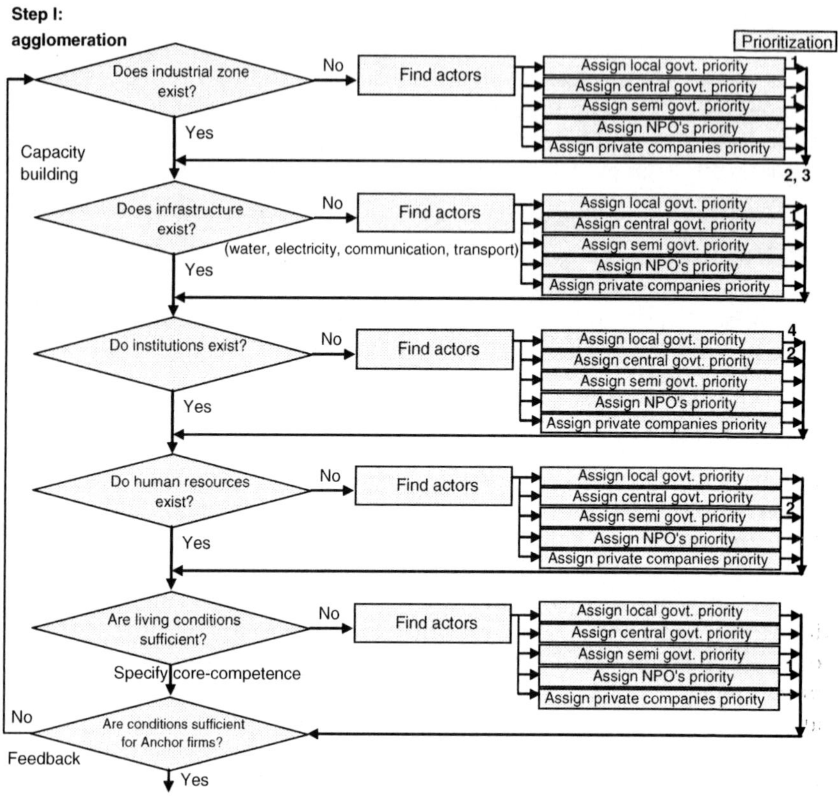

Figure 1.2 Flowchart approach: Step I. agglomeration
Source: Kuchiki (2007).

Following identification of the actors responsible for forming industrial zones, we return to the "main stream" of the flowchart to examine the components of capacity building. Following Figure 1.2, we determine whether water supply is sufficient for the industrial zones. If we verify the sufficiency of water supply, we proceed down along the flowchart to power supply, communication and transportation. An example of the importance of such infrastructure is the shortage of power supply in 2004 and 2005 in China's Guangdong Province, a serious concern for that cluster. The central government of China and the local government of Guangdong Province were the actors that moved to increase power supply. The central government drew up plans for building a nuclear power plant, constructed dams, and distributed electricity by estimating the total supply and demand of electricity in China. The local government of Guangdong Province planned to generate and distribute electricity. The local government also attempted to alleviate the region's shortage of railways, subways and highways by building additional capacity in 2005.

After physical infrastructure is examined, we look at whether institutions have been developed and are ready to function. The central government must institutionalize national tax systems, while local governments must institutionalize local tax systems. It is well known that one-stop services for investment procedures are crucial to success in inviting foreign investors.

Concerning human resource development, during the early stages of a country's industrialization, a pool of highly literate unskilled labor is crucial to inviting foreign investors desiring to employ cheap labor. After industrialization progresses, an industrial cluster sometimes faces a shortage of skilled labor, requiring the presence of universities and on-the-job training centers for innovation in order to sustain the development of the cluster.

Superior living conditions are crucial to inviting foreign investors. Staff members of foreign investor firms will have incentives to work hard if they can enjoy a superior quality of life in their region. Sufficient housing, schools, hospitals, and other such facilities must be developed in order to successfully invite anchor firms to a region.

We explain the priorities of each cluster actor in Figure 1.2. Local governments play a vital role in constructing industrial zones, supplying electricity, providing water and wastewater services, facilitating transport, and developing institutions. The central government's main functions are to build institutions and construct nuclear power stations, with both of these at about an equal priority level.

Next, Figure 1.3 clarifies the nature of a cluster by dividing step I, the process of industrial agglomeration, from step II, which involves the innovative activities of the agglomerated firms. Here, we define an industrial cluster as consisting of both "industrial agglomeration" (step I) and "innovative activities" (step II). Industrial agglomeration means that related firms are located in the same region; innovative activities refer to active innovation efforts by the agglomerated firms. Figure 1.3 shows the industrial agglomeration process of step I. Step I consists of industrial zones, capacity building (I), anchor firms, and related firms. Step II is a process of innovation, whose elements are universities and research institutes, capacity building (II), and anchor persons. Capacity Building (II) is different from Capacity Building (I). The conditions for innovation hold if this second type of capacity is built up and if anchor persons come.

We now explain the analysis manuals of the flowchart approach, presented in Tables 1.1, 1.2 and 1.3. The first row of Table 1.1 specifies a city as the core of a cluster. The second row lists industries in which clusters may form, such as the automobile industry, the electronics industry, the information industry, and the biotechnology industry. The third row confirms whether there are industrial zones as saucers to invite investors, whether the cluster's products are intended for domestic or export markets, when the

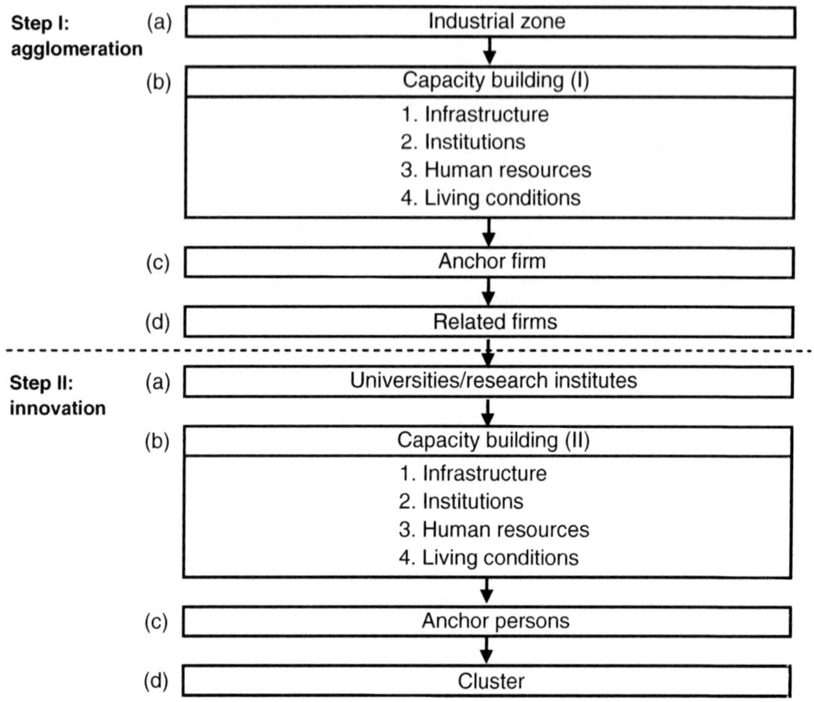

Figure 1.3 Flowchart approach to industrial cluster policy
Source: Kuchiki (2007).

Table 1.1 Format of the analysis I (check list)

(1) City: Hanoi, Tianjin, Penang, Guangzhou, etc.			
(2) Industry: automobile, eletronics, information technology, software, biology, etc. : **number of firms** related to an anchor firm			
(3) (a) Industrial zones	Market	Established year	Economic agents
	Domestic or export		Public, semipublic or private
(c) Anchor firm			
(4) Case: success or failure: **core competence**			

Source: author.

industrial zones were established, and which actors established the zones. Again, the possible actors include the central government, local governments, semi-governmental actors such as authorities, and private companies.

Table 1.2 provides a rubric for determining whether the capacity represented by infrastructure, institutions, human resources, and living

Table 1.2 Format of the analysis II: capacity building

			Actors					
			Sufficient or not	**Local gov.**	**Central gov.**	**Semi-gov.**	**NPOs**	**Private companies**
	IZ or EPZ							
Capacity	1. Infrastructure	(1) Water	–					
		(2) Electricity	–					
		(3) Communication	–					
		(4) Transport	–					
	2. Institutions	(1) One-stop services	–					
		(2) Deregulation	–					
		(3) Preferential treatments (tax incentives, etc.)	–					
		(4) Laws and regulations (bankruptcy laws and intellectual property right)	–					
	3. Human resource	(1) Unskilled labor	–					
		(2) Skilled labor	–					
		(3) Professionals	–					
	4. Living conditions	(1) Housing	–					
		(2) International schools	–					
		(3) Hospitals	–					
		(4) Entertainment	–					

Source: author.

Table 1.3 Format of the analysis III: related firms

Related firms	Established year	Products

Source: author.

conditions is sufficient for the success of the industrial cluster policy. Infrastructure includes water, electricity, communication and transportation. Institutions include one-stop services, deregulation and preferential treatments such as tax incentives, laws and regulations. Human resources include unskilled labor, skilled labor, and professionals. Living conditions include housing, schools, hospitals, shopping, and entertainment.

If a deficiency is identified, it is necessary to identify the actors responsible for addressing those shortages. These actors will facilitate the success of the industrial cluster policy by meeting the capacity shortages, as discussed earlier.

Once an anchor firm decides to invest in an industrial zone, that company's related firms decide whether to move into the industrial zone. Their decisions depend on the production volume of the anchor firm – in other words, on economies of scale. The related firms first estimate the costs and benefits of investment in the industrial zone and then make their decision accordingly. We compiled a list of examples of related firms in Table 1.3.

We should note that this prototype model cannot be applied to the biotechnology industry or the information technology industry, partly because those industries do not use a large number of parts and components in the manufacturing process and thus are not subject to economies of scale.

3. Summary of this book

This book consists of 9 chapters, each examining a theoretical aspect or empirical case study of the clustering process, namely, the first two chapters provide theoretical foundations of the flowchart approach, while the rest of chapters present current situations of clustering and policy in different economies as well as empirical studies. Let us introduce each chapter in more detail.

Masahisa Fujita's "Formation and Growth of Economic Agglomerations and Industrial Clusters: a Theoretical Framework from the Viewpoint of Spatial Economics" presents a microeconomic framework for understanding the formation and growth of industrial clusters, drawing on the theory of spatial economics. Following the work of Ann Markusen, he presents a typology of industrial agglomerations, describing four basic types of agglomeration. In reality, most agglomerations exhibit a mix of the characteristics of the four types, and evolve gradually over time. Thus, the author presents a general framework for understanding the processes by which a wide variety of economic agglomerations and industrial clusters form and evolve, drawing on spatial economic models. The author explains in detail the economic mechanisms that give rise to clustering forces, focusing especially on endogenous models. Next, he examines the general effects of decreasing transport costs on the spatial distribution of economic activities; this is significant because the main engine for the recent dynamics of the global economy has been the steady reduction of (broadly defined) transport costs. Finally, the author explains the basic economic reasons for the formation of multinational firms (MNFs), as the rapid proliferation of these firms has been a prominent phenomenon in the current wave of globalization.

"The Flowchart Approach to Industrial Cluster Policy: Guangzhou's Automobile Industry Cluster," by Akifumi Kuchiki and Hiroyuki Tsukada, constructs a prototype model of the flowchart approach to industrial cluster policy and applies it to Guangzhou's automobile industry cluster, demonstrating the effects of the anchor firms Honda, Nissan, and Toyota on the cluster. The authors obtain two important conclusions. First, their flowchart approach is effective in analyzing the implementation and effects of Guangzhou's automobile industry cluster policy. Second, the local government of Guangzhou Municipality plays a crucial role in the success of the policy, by promoting joint ventures between Guangzhou Automobile and foreign firms (including the three Japanese firms mentioned above). The mayor of Guangzhou Municipality, whose job promotion is linked to regional economic performance, has an incentive to target industrial clustering and implement cluster policy.

In "Industrial Clusters in the Austin Area: The Austin Technopolis Case Study," Jobaid Kabir analyzes the robustness of the framework established by Dr. Kuchiki, and presents a broad overview of the process and causes of cluster formation in Austin in the U.S. He shows that, while Austin's high technology cluster development does not precisely follow Kuchiki's flowchart framework, there are significant similarities. The author presents a "revised flowchart" to describe Austin's cluster development. In this revised flowchart, the anchor firms first conduct a nationwide search for a location. Cities and states then compete with each other to meet the anchor firm's needs, providing plans and designs for capacity building in the area. Once the anchor firm selects a location, major capacity building activities such as infrastructure construction begin, and related firms then move into the area to support the anchor firms. In Austin's case, key individuals and visionaries played a significant role in facilitating interaction among various stakeholders in the service of a common goal.

Aya Okada and N.S. Siddharthan, in "Automobile Clusters in India: Evidence from Chennai and the National Capital Region," analyze the spatial patterns of industrial agglomeration of the consumer electronics, electronics and hardware, and drugs and pharmaceuticals industries. They show that these industries are all geographically concentrated in the National Capital Region (Delhi and parts of Haryana and Uttar Pradesh adjacent to Delhi), Maharashtra, Gujarat, Tamil Nadu and Karnataka. In particular, firms in the Indian auto industry are mostly agglomerated in three main clusters: Chennai (Tamil Nadu), Pune-Mumbai (Maharashtra), and the NCR. Contrasting the situations of the Chennai and NCR clusters, the authors demonstrate the presence of inter-cluster variations in the patterns of auto cluster formation, and show that these differences are partly explained by the historical and policy conditions under which firms, particularly anchor firms, operate in these regions. Nevertheless, these two clusters share some common features: the creation of industrial estates by state governments, a large pool of well-educated workers, and the strong influence of government policy (especially industrial licensing and location policies). The authors' econometric analyses confirm that location in a cluster positively influences the performance of auto component firms, with those belonging to a cluster performing significantly better.

"The Process and Factors of Industrial Cluster Formation: A Flowchart Approach to Industrial Cluster Policy in Japan," by Kentaro Yoshida, provides an overview of the current condition of various types of industrial clusters in Japan whose formation was influenced by government policy, comparing and contrasting the factors that led to these clusters' formation. The paper then presents the results of a mail survey administered to public institutions responsible for cluster promotion, and uses these results to quantitatively test a hypothesis concerning the process and causes of industrial cluster formation. The mail survey demonstrates that anchor firms, related firms, universities, and research institutes are thought to have been effective factors in forming industrial clusters in Japan. The author concludes that there is a striking trend toward the development of knowledge-intensive economies, wherein there is a shift away from industrial agglomeration strategies toward innovation-based strategies. This shift is evident in the infrastructure building (incubation), system support, and human resources training being conducted as part of industrial cluster policy across Japan. the author stresses the necessity for the presence of cooperation promotion agencies, which enhance industrial cluster formation.

"The Evolution of the High-Tech Electronics Cluster in Guadalajara, Mexico," by Yoshiaki Hisamatsu, examines how and why Guadalajara has risen to prominence as "the Mexican Silicon Valley." The cluster is largely comprised of foreign companies, but there are some domestic firms as well. According to Hisamatsu, these major companies chose Guadalajara as a production site due to a variety of factors, chief among which were macroeconomic contingency (massive exchange rate devaluation in 1994), market proximity to the U.S., a relatively stable labor market environment, skilled human resources trained by local universities, history (prior investment), and institutional support. He also highlights the contributions of individuals who helped the cluster to exploit the above-mentioned locational advantages – local professional managers who promote the cluster through the local business association.

Examining the Mexican Silicon Valley through the lens of the flowchart approach, the author claims that the local business association (CANIETI) and its spin-off organization (CADELEC), functioning as quasi-public actors, provide the cluster with various coordination and information services that can be interpreted as quasi-public goods. These quasi-public actors establish a valuable process of trial and error that informs all businesses in the cluster. The author predicts that, through this collective learning process, the cluster will be able to continue to grow in the competitive global electronics market.

In "An Empirical Examination of the Flowchart Approach to Industrial Clustering: Case Study of Greater Bangkok, Thailand," Masatsugu Tsuji and Yasushi Ueki analyze agglomeration in Greater Bangkok in order to empirically verify the applicability of the flowchart approach. Based on data obtained by mail surveys given to firms located in Bangkok, the authors construct three models of clustering, utilizing probit estimations to isolate the impact of various clustering factors. The models take the year in which companies established their Bangkok offices as the dependent variable, with company attributes and clustering factors as the explanatory variables. Some of these clustering factors are predicted by the flowchart approach: namely, domestic demand, export, inter-firm relationships, the degree of pre-existing agglomeration, and capacity, which includes infrastructure, institutions, human resources, and living conditions. The authors obtain the following results from their estimation: Anchor firms entered the cluster first, followed by smaller firms. Capacity, in the form of "transportation infrastructure such as roads and ports" and "government policy regarding tax and investment," attracted these first-movers. The domestic market had a positive effect on agglomeration during the period of 1987–1994. Export, however, had no significance. Related and affiliated firms mainly came to Bangkok after the anchor firms. These estimation results largely support the flowchart approach.

4. Conclusions

Taken together, these studies capture the essential features of the industrial agglomerations currently forming in East Asia, and show how clusters in other regions share some key characteristics with their East Asian counterparts. The flowchart approach, which we show to be widely applicable to clusters in various industries, can therefore be used to construct industrial cluster policies. By identifying and prioritizing the factors required for agglomeration formation and cluster development, the flowchart approach offers a growth strategy for regional as well as national economies. The two major conclusions of this volume can be stated as:

1. Capacity building is a key factor to the success of the industrial cluster policy;
2. Capacity building should be targeted to invite anchor firms.

Although this volume focuses on the flowchart approach as an explanation of industrial agglomeration, it should be noted that the approach cannot explain all phenomena. Further research is required to identify the range of industries and industrial development levels for which the flowchart approach is most applicable. Moreover, some East Asian economies are reaching the innovation stage, progressing beyond simple production bases, and endogenous innovation and R&D activities are proliferating in these areas. These are natural outcomes of agglomeration, emphasized by Fujita and Thisse [2002]. Agglomerations enable and enhance exchanges of information, know-how, and even tacit knowledge, all of which promote the endogenous flow of innovation or R&D. The flowchart approach must be modified in the future in order to explain these processes.

Bibliography

Baldwin, R. *Economic Geography and Public Policy.* Trenton, New Jersey: Princeton University Press, 2003.

Fujita, M. "Significance and Issues of Japanese Industrial Cluster Policy from the Point of Spatial Economics." In Y. Ishikura (ed.), *Japan's Industrial Cluster Strategy* (in Japanese) Tokyo: Yuhikaku, 2003.

Fujita, M. and J.-F. Thisse. *Economics of Agglomeration.* Cambridge: Cambridge University Press, 2002.

Kuchiki, A. "Agglomeration of Exporting Firms in Industrial Zones in Northern Vietnam." In Tsuji, Giovannetti, and Kagami (eds), *Industrial Agglomeration and New Technologies*. Cheltenham: Edward Elgar, 2007.

Kuchiki, A. "A Flowchart Approach." In A. Kuchiki and M. Tsuji (eds), *Industrial Clusters in Asia*. London: Palgrave Macmillan, 2005.

Kuchiki, A. *Theory of Asia's Industrial Clusters (in Japanese).* Tokyo: Hayama, 2007.

Kuchiki, A., and M. Tsuji, (eds), *Industrial Clusters in Asia: Analyses of their Competition and Cooperation*. London: Palgrave Macmillan, 2005.

Porter, M. E. *The Competitive Advantage of Nations*. New York: Free Press, 1998.

Tsuji, M., E. Giovannetti and M. Kagami (eds), *Industrial Agglomeration and New Technologies: A Global Perspective*. Cheltenham: Edward Elgar, 2007.

Tsuji, M. and W. Quan. "Chinese Automotive and Parts Industries." In Kuchiki and Tsuji (eds), *Industrial Clusters in Asia*. London: Palgrave Macmillan, 2007: 200–221.

2

Formation and Growth of Economic Agglomerations and Industrial Clusters: A Theoretical Framework from the Viewpoint of Spatial Economics

Masahisa Fujita

1. Introduction

Drawing on the framework of spatial economics, this chapter presents a microeconomic framework for understanding the formation and growth of economic agglomerations and industrial clusters. Spatial economics, a field created in the early 1990s, represents a new branch of economics that aims to explain the formation and growth of various forms of economic agglomeration in geographical space, using a general equilibrium framework combined with an evolutionary approach. Spatial economics is often called "the new economic geography," and the two terms are often used interchangeably throughout this chapter.[1] Although the chapter's main focus is on the theory used to explain industrial agglomerations and clusters, we also consider the difference between various forms or types of economic agglomerations. Given that most industrial agglomerations in the real world are embedded in a larger agglomeration such as a city or a region, a broader viewpoint is helpful to achieve a comprehensive understanding of the formation and growth of a specific industrial agglomeration or cluster.

Agglomeration, or the clustering of economic activity, occurs at many geographical levels and varies greatly in its composition. At one extreme lies the core-periphery structure at the global scale. In 2000, for example, the NAFTA countries yielded 35% of world GDP, the EU(15 countries) 25%, and East Asia 23%; thus, 83% of world GDP was concentrated in three regions. In 1980, however the shares were 27% for NAFTA, 29% for the EU, and 14% for East Asia- the three regions together yielding only 70%. Hence, world

GDP has been increasingly concentrating in these three regions. Furthermore, Hall and Jones (1999) observed that high-income nations are clustered in small cores in the Northern Hemisphere, and that worker productivity steadily declines with distance from these cores.

Strong regional disparities within the same country imply the existence of agglomerations at another spatial scale. In France, for example, the Ile-de-France (the metropolitan area of Paris), which accounts for 2.2% of France's area and 18.9% of its population, produces 30% of the country's GDP.

Regional agglomeration is also demonstrated by the large variety among cities, as shown by the stability of the urban hierarchy within most countries (Mori, Nishikimi and Smith 2003). Cities themselves may be specialized in a small number of industries, as are many medium-size American cities (Henderson 1997). However, large metropolises like New York and Tokyo are highly diversified in that they nest many industries that are not related through direct linkages (Fujita, Henderson, Kanemoto and Mori 2004). Industrial districts involving firms with strong technological or informational linkages, or both (e.g. Silicon Valley, or Italian districts with more traditional economies), as well as factory towns (e.g., Toyota City in Japan and Hershey in the US), manifest various types of local specialization. Therefore, it appears that agglomerations of highly diverse size and structure exist at both the regional and urban levels.

At the most detailed end of the spectrum, agglomeration arises in the form of large commercial districts that exist within a city itself (such as Soho in London, Montparnasse in Paris, or Ginza in Tokyo). At the very smallest level, restaurants, movie theaters, or shops selling similar products are often clustered within the same neighborhood; clustering may even take the form of a large shopping mall.

It is also important to realize that all these different types and levels of agglomeration are embedded in a larger economy, together forming a complex system. A wide understanding of these phenomena is critical for the design of effective urban and regional policies.

To some extent, economic activities are spatially concentrated because of dissimilarities in such natural features as rivers, harbors and mineral deposits, called "first nature." However, the impact of first nature on the spatial distribution of economic activities is not difficult to explain within the traditional economic theory of the competitive paradigm. Thus, the focus of this chapter is on the economic modelling of endogenous mechanisms (or *second nature*) that lead to agglomeration. Furthermore, an economic model of agglomeration should explain both concentration and dispersion: in France, for example, why so many people live in Île-de-France and also why so many others do not; or, in the New York Metropolitan Area, why so many people work in Manhattan and also why so many others do not (Fujita and Krugman 2004). To put it in another way, an economic model of agglomeration is expected to provide a general equilibrium story about the centripetal forces that pull economic activities together and the centrifugal forces that push them apart, explaining these in terms of the trade-offs between various forms of increasing returns and different types of mobility costs.

Spatial economics provides a general analytical framework for the explanation of the formation and growth of various kinds of economic agglomerations. To date, spatial economics remains the only general equilibrium framework in which the location of economic agglomerations is determined explicitly through a micro-economic mechanism.

Although the main objective of this chapter is to present a comprehensive theoretical framework for understanding industrial clusters and agglomerations from the viewpoint of spatial economics, we also aim to reconsider here the *industrial cluster policy,* first formulated by Michael Porter, from the viewpoint of spatial economics. Since the early 1990s, not only economists and economic geographers but also management scholars and policy makers have been paying increasing attention to the role of industrial clustering in promoting the development of urban and regional economies. This is partly due to the extensive documentation of success stories of industrial agglomerations and clusters in various countries around the world. Famous example include the study of the Third Italy by Piore and Sabel (1984) and the comparison of Silicon Valley and Route 128 by Saxenian (1994). Furthermore, since the extensive work of Porter (1998) on industrial clusters, many policy makers in national and regional governments in both developed and developing countries have begun adopting "industrial cluster policy" as an important strategy for economic development.

Reconsidering industrial cluster policy from the viewpoint of spatial economics is expected to be useful in several respects. First, given Porter's theory of clustering based on management theory, spatial economics may provide a useful alternative framework for analyzing policy. As is explained in detail in Section 3, spatial economics represents a new branch of economics that aims to explain the formation and location of a large variety of economic agglomerations in geographical space. Since both Porter's clustering theory and spatial economics focus on industrial agglomerations and clusters, but are based on different theoretical frameworks, the two complement each other. Second, while Porter focuses specifically on a class of industrial agglomerations in which innovation is continuous, spatial economics examines a broader class of economic agglomerations, and can therefore help to reconsider Porter's cluster theory from a broader perspective without limiting ourselves to the case of innovative agglomerations. Finally, in reality, most existing industrial clusters, including planned ones, are embedded in existing cities or metropolitan areas; thus, the growth of an industrial cluster is often inseparable from the growth of the city in which it is embedded. Therefore, given that one of the main concerns of spatial economics is the formation and growth of cities, spatial economics can complement Porter's clustering theory from the viewpoint of urban development. For these reasons, we can expect the study of industrial clusters and agglomerations to be more fruitful if we can effectively combine Porter's clustering theory with spatial economics.

In the rest of this paper, we usually use the terms of *industrial agglomerations* and *industrial clusters* synonymously. However, when we focus on Porter's type of industrial agglomerations (in which innovation is continuous), we call these "Porter-type clusters."

An outline of the rest of this chapter is as follows: In section 2, following Markusen (1996), we present a typology of four basic types of industrial agglomerations. In reality, many agglomerations exhibit a mix of the characteristics of the four types, and evolve gradually over time. Thus, in Section 3, we present a general framework for understanding the formation and the evolutionary progression of economic agglomerations and industrial clusters, from the viewpoint of spatial economics. In this framework, the principal concern is how to explain the forces behind the formation of a large variety of spatial agglomerations. Hence, in Section 4, we explain in detail the economic mechanisms that yield agglomeration forces, focusing on endogenous agglomeration forces in particular. Next, given that the main engine for the recent dynamics of the global economy is the steady reduction of transport costs (broadly defined), in Section 5 we examine the general effects of decreasing transport costs on the spatial distribution of economic activities. In the recent process of globalization, a prominent phenomenon is a rapid increase of multinational firms (MNFs). Thus, in Section 6 we explain the basic economic reasons for the formation of MNFs. Finally, Section 7 concludes the paper.

2. Typology of industrial agglomeration

In order to develop a comprehensive theory of industrial agglomerations, which is the main objective of this chapter,it is useful first to consider a classification of the possible types of industrial agglomerations. In this respect, Markusen (1996) presents a useful typology of industrial agglomerations (or "industrial districts" in Markusen's terminology).[2] According to Markusen (1996), an industrial agglomeration is a large, and spatially delimited area of trade-oriented economic activity that has a distinctive economic specialization, either resource, manufacturing, or servicesrelated.

Based on intensive inductive research in several countries, Markusen (1996) offers the following four distinct types of industrial agglomeration:

(A) Marshallian Type
(B) Hub-and-Spoke Type
(C) Satellite Platform Type
(D) State-Anchored Type

In Figure 2.1, types (A), (B), and (C) are depicted schematically.

The hypothesized features of each type are briefly summarized below.[3]

A. *Marshallian type* industrial agglomerations are characterized by a business structure dominated by small, locally owned firms, where scale economies of individual firms are relatively small, while substantial intra-agglomeration trade takes place locally among buyers and suppliers. *Italian industrial districts* represent a modern version of the Marshallian type. The Silicon Valley IT cluster can be considered to represent an advanced version of this type.

B. *Hub-and-Spoke* industrial agglomerations are characterized by a business structure dominated by one or several large, vertically integrated firms surrounded by suppliers. In this type of agglomeration, core firms are embedded non-locally, and maintain substantial links to suppliers and competitors outside of the agglomeration, while the scale economies of core firms are relatively high. The agglomeration of the automobile industry centered around Toyota in Aichi Prefecture represents a classic example of this type.

C. *Satellite platform* industrial agglomerations are characterized by the business structure dominated by large, externally owned and headquartered firms, where scale economies of core firms are moderate to high, and minimal intra-agglomeration trade exists among buyers and sellers. The IT cluster in Japan's Kyushu Region represents a good example of this type.

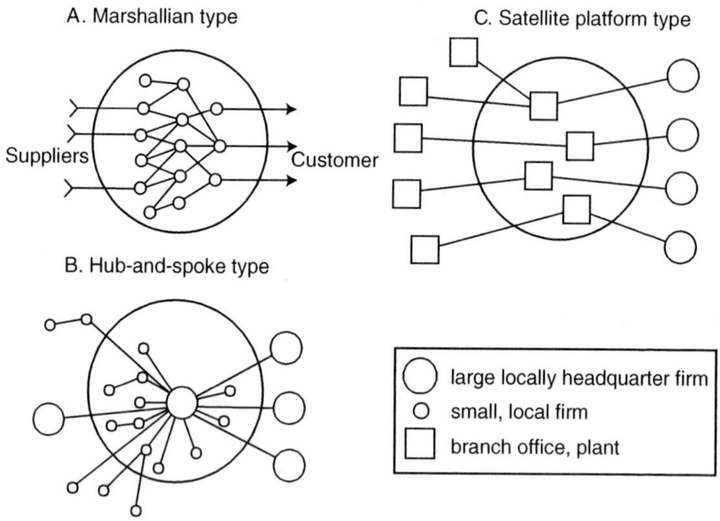

Figure 2.1 Three basic types of industrial agglomerations
Source: Markusen (1996).

D. A state-anchored industrial agglomeration is characterized by one or several large government institutions such as military bases, state or national capitals, or large public universities, surrounded by their suppliers and customers. Scale economies are relatively high in public-sector activities, and substantial intra-agglomeration trade takes place among dominant institutions and their suppliers. This agglomeration type tends to look much like the hub-and-spoke type agglomeration, (B), in Figure 2.1. The initial stage of many of the fastest-growing industrial agglomerations in the US (such as Santa Fe, San Diego, Silicon Valley, and many state university-dominated cities) is characterized by this type of agglomeration.

Although types A through D above represent idealized types of hypothesized industrial agglomerations, in reality most industrial agglomerations exhibit a mix of the characteristics of the various types, as well as evolving over time. In particular, large metropolitan areas typically exhibit elements of all four types. Markusen (1996) notes Silicon Valley as a typical case in which elements of all four types are strongly present. As another example, take the agglomeration of the automobile industry in the Bangkok Metropolitan Area (BMA), which was started about four decades ago by Toyota in the simplest form of type C (Satellite Platform), but soon evolved to type B (Hub-and-Spoke) with the growth of many local suppliers, and now looks like a huge Marshallian type agglomeration centered around a dozen major international automobile producers operating there. Furthermore, the automobile agglomeration in Bangkok is embedded in the huge BMA, which is home to 6 million people, as well as most categories of manufacturing and service industries and government activities.

In the following section, we present our general framework for understanding the formation and the evolutionary process of economic agglomerations and industrial clusters, which proceeds from the standpoint of spatial economics.

3. Spatial economics – its birth and basic framework

In this section, we explain a brief history of the development of spatial economics, as well as its basic framework. Spatial economics is often called the "new economic geography." In order to emphasize that we are mainly concerned with recent developments in spatial economics, in the rest of this section we refer to the theory as the new economic geography.

3.1. The birth of the new economic geography

Until very recently, there were three distinct fields of economics that each dealt with the spatial aspects of human activity. These were: *urban economics*, which focuses on cities; *regional economics*, which, as its name implies, is the study the of so-called "regions," and *international trade theory*, which concentrates on international industrial specialization and trade. However, with the rapid development towards a borderless global economy over the last two decades, it has become increasingly clear that none of the traditional theories of urban economics, regional economics, or international trade would prove adequate or appropriate for the study of the dynamics of the spatial economies that are now taking place in almost all parts of the world.

For example, regarding the formation of the EU in 1993, an obvious question of interest was (and still is), what will be the new spatial distribution of economic activity in the EU? For that matter, what will be the new economic geography of the expanded 25-country EU? As market integration dissolves economic barriers between nations, it is apparent that traditional international trade theory, which depended crucially on the concept of national boundaries, does not provide an appropriate framework for the analysis of such questions. Furthermore, traditional urban economics treated cities as "floating islands" and did not deal with the issues involving the location of cities, an aspect of analysis that is crucially important for the study of the borderless EU. Furthermore, the concept of "regions" used in traditional regional economics is quite vague, at best.

Given the limitations of existing spatial economic theories, a strong demand arose in the late 1980s for a new, more general theory of spatial economy that would be appropriate for the study of an increasingly borderless world economy. And as Professor Paul Samuelson once noted, "wherever there is demand, there will be supply."

Indeed, as is well-known, there has been since about 1990 a renaissance of theoretical and empirical work on economic geography. Among others, the pioneering work of Paul Krugman (1991a, 1991b) on the core-periphery model has triggered a new flow of interesting contributions to economic geography or spatial economics. Following soon after this pioneering model, Anthony J. Venables and Paul Krugman together developed several interesting models of international trade and industrial agglomeration/specialization, while Krugman and Fujita worked together on the evolution of cities and urban systems. At the same time, an increasing number of young economists all over the world joined in the efforts of developing a new approach for the study of economic geography.

Together, these economists' work was labelled "the new economic geography," and has quickly emerged as one of the most exciting areas of contemporary economics. A comprehensive manifestation of this approach is seen in the book, *The Spatial Economy: Cities, Regions and International Trade* (1999, MIT Press), authored by Fujita, Krugman and Venables.[4] The new economic geography is expected to play a major role in economic analyses of the increasingly borderless global economy.

3.2. The basic framework of the new economic geography

The defining issue of the new economic geography is how to explain the formation of a large variety of economic agglomerations (concentrations of activity) in geographical space. Agglomeration occurs at many geographical levels and is composed of a variety of elements. For example, one type of agglomeration arises when small shops and restaurants are clustered in a neighborhood. Other types of agglomeration can be found in the formation of cities of many different sizes, ranging from Tokyo (with over 30 million people), and Bangkok (with some 6 million), to countless smaller ones. Other types of agglomeration can be seen in the emergence of a variety of specialized industrial districts, such as the concentration of electronics industry in the famous Silicon Valley in the US and in Kuala Lumpur in Malaysia. Furthermore, spatial agglomeration on a greater geographical scale is manifested in the existence of strong regional disparities within each country. At the largest extreme of the spectrum lies the core-periphery structure of the global economy, corresponding to the "North-South Dualism." It is also important to notice that all these different types and levels of agglomeration are embedded in a larger economy, together forming an extremely complex system.

The traditional theory of international trade and regional economics provides a useful framework to explain the dispersion of economic activity based on factor price differentials. However, it cannot explain the agglomeration of economic activity in a borderless economy. If we use an analogy in terms of geology, the traditional trade theory provides a useful framework to explain how mountains will become flatter and flatter through erosion by water flowing from the top to the bottom; namely, the convergence process. But it does not explain why we have mountains in the first place. Table 2.1 contrasts the framework of this new school with that of the traditional (neoclassical) international/regional economics.

The hallmark of the new economic geography is the presentation of a unified approach to modelling a spatial economy characterized by a large variety of economic agglomerations, one that emphasizes the three-way interaction among increasing returns, transport costs (broadly defined), and the movement of production factors, in which a general equilibrium model is combined with nonlinear dynamics and an evolutionary approach for equilibrium selection. Figure 2.2 represents the basic conceptual framework of the new economic geography.

The observed spatial configuration of economic activities is thought to be the outcome of a process involving two opposing types of forces, that is, *agglomeration* (or *centripetal*) *forces* and *dispersion* (or *centrifugal*) *forces*. A complicated balance of these two opposing forces leads to the emergence of

Table 2.1 Contrasting the two theoretical frameworks

Traditional international/ regional economics	**New economic geography**
Constant returns	Increasing returns
Perfect competition	Imperfect competition
Uneven distribution of resources (first nature)	Endogenous agglomeration forces (second nature)
Borders	Transport costs
Static/long run equilibrium	Self-organization/ evolutionary

Source: author.

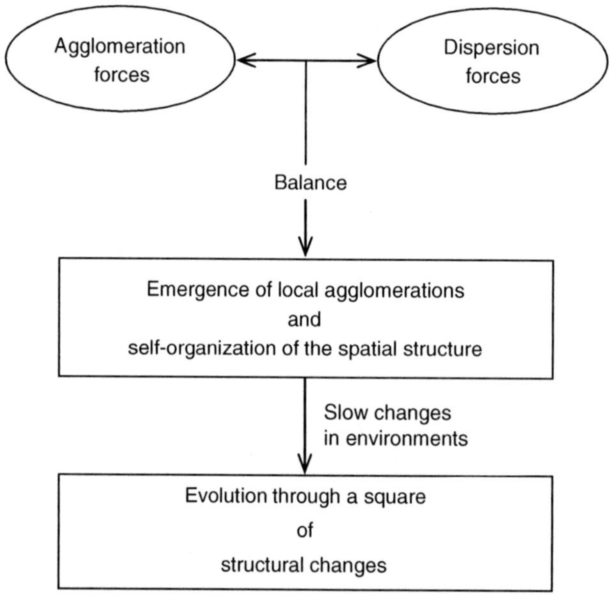

Figure 2.2 The basic framework of the new economic geography
Source: author.

a variety of local agglomerations of economic activity, and the spatial structure of the entire economy is self-organized. With gradual changes in the technological and socio-economic environments, the spatial system of the economy experiences a sequence of structural changes, evolving toward an increasingly complex system.

4. The nature of agglomeration forces

In this framework, then, the first two questions of obvious importance are:

Question 1: how can we explain agglomeration forces?
Question 2: how can we explain dispersion forces?

The answer to Question 2 is rather easy, for the concentration of economic activities at a location will naturally increase factor prices (such as land prices and wage rates) and induce congestion effects (such as traffic congestion and air pollution), which can all be readily explained by the traditional economic theory. Thus, the principal concern of the new economic geography is Question 1, i.e., how to explain the agglomeration forces.

In spatial economics, as depicted in Figure 2.3, agglomeration forces are thought to be generated from the combination of three basic elements: *first nature, second nature,* and *catalyst.*

In Figure 2.3, *first nature* represents, for example, the presence of natural ports and reveres, natural resources such as mineral deposits and water, and natural environments such as pleasant weather, greenery, or beautiful beaches. *Second nature* refers to the endogenous agglomeration forces generated within the economic system, which are explained in detail below. Catalysts represent the various factors initiating and fostering the growth of

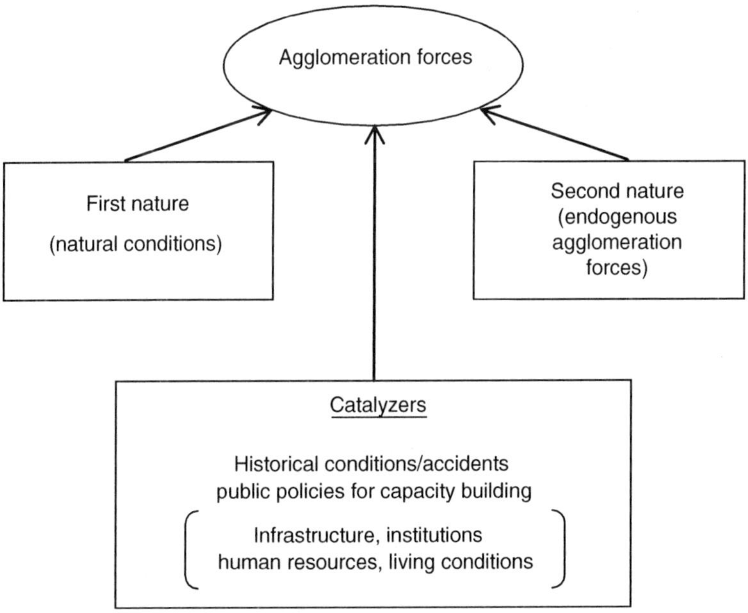

Figure 2.3 Generation of agglomeration forces

Source: author.

a specific agglomeration, such as historical conditions, accidents, and various public policies (known as "capacity building").[5] To some extent, economic activities are spatially concentrated because of dissimilarities in first nature. However, the impact of first nature on the spatial distribution of economic activities is not difficult to explain within traditional economic theories based on the competitive paradigm. Furthermore, the theory of spatial economics emphasizes the self-organizing aspect of economic agglomerations. Thus, in the rest of this section, we focus on the economic mechanisms that give rise to endogenous agglomeration forces.

Figure 2.4 presents the general principle that lies behind these economic mechanisms.

This figure represents the idea that under the presence of sufficient heterogeneity (i.e., differentiation) in goods (including services) or workers, the three-way interaction among increasing returns (at the individual firm level), transport costs (broadly defined), and migration of workers (who are also consumers) creates a circle of causation that leads to the agglomeration of both consumers (users) and suppliers of these goods or services. Here, the first key element is the *heterogeneity* in goods; when goods are sufficiently differentiated from each other, their suppliers can locate in proximity without involving severe price competition, while consumers (or users) can enjoy

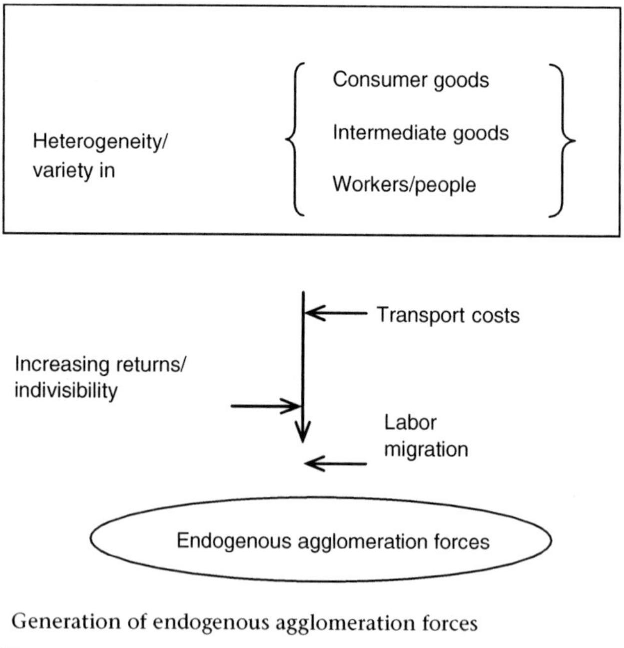

Figure 2.4 Generation of endogenous agglomeration forces
Source: author.

the complementarity of such heterogeneous goods by locating close to their suppliers. The second key element is *increasing returns* at the firm level and the indivisibility of a human being. In fact, without scale economies at the firm level, there is no need to concentrate the production of each good at the same location. That is, without scale economies, the economy will degenerate into "backyard capitalism," in which each household or small group produces most items itself. The third key element is *transport costs*. Indeed, without transport costs (broadly defined), "location" does not matter. The presence of transport costs gives rise to the "home market effect" for suppliers located near a large market. Finally, the *migration of workers* (=consumers) is a prerequisite for the agglomeration of workers and firms together.

Focusing on the heterogeneity in consumer goods, Figure 2.6 demonstrates the circle of causation that leads to the agglomeration of the producers of goods and their consumers. Starting with the bottom box, for example, suppose that a large variety of consumer goods is produced in a city. Because of the transport costs, this variety of goods can be purchased at lower prices in that city than in places far away from the city. Thus, given a nominal wage in the city, consumers' taste for variety leads to a rise in the real income of workers in that city. This, in turn, induces more workers to migrate to the city. The resulting increase in the number of consumers (=workers) thus creates a greater demand for goods in the city. Due to the home market effect (i.e., the benefits of locating near a large market), even more firms, supplying new varieties of goods, choose to locate in the city. This leads to the availability of an even greater variety of goods in the city, which begins the cycle anew. Thus, as depicted in Figure 2.5, a *circle of causation* of the agglomeration of firms and workers in the city is created through *forward linkages* (the supply of a greater variety of goods increases

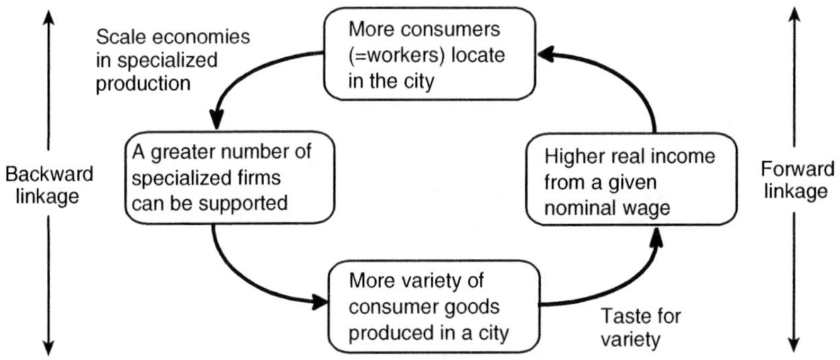

Figure 2.5 Circle of causality in spatial agglomeration of consumer-goods producers and workers (=consumers)

Source: author.

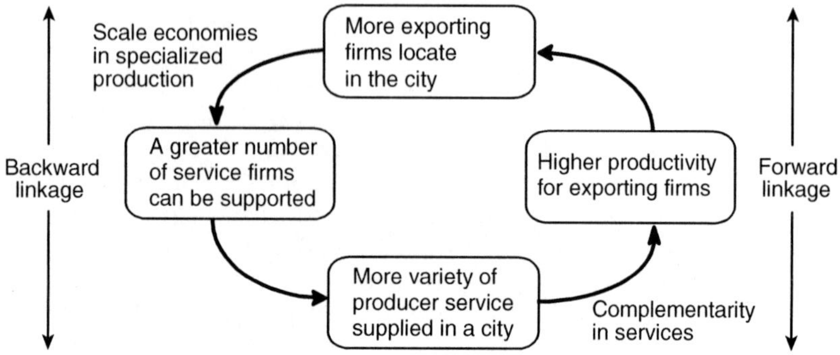

Figure 2.6 Circle of causality in spatial agglomeration of final-good producers (=exporting firms) and producer-services
Source: author.

the workers' real income) and *backward linkages* (a greater number of consumers attract more firms). In other words, through the *pecuniary externalities* caused by these linkage effects, scale economies at the individual firm level are transformed into increasing returns at the level of the entire city.

Likewise, taking producer-services as an example of an intermediate good, Figure 2.6 explains the relationship between the agglomeration of final-good producers and the suppliers of a large variety of producer services in a city. This type of agglomeration force, caused by product variety in producer services (more generally, intermediate goods), can partly explain the concentration of high-technology firms or heavy industries (e.g., Silicon Valley and Toyota City) or business-service firms (e.g., New York, Tokyo and Hong Kong).

Finally, we examine the effects arising from the diversity of people – the effects of specialized workers called "brain-workers." Figure 2.7 depicts the circle of causation leading to the agglomeration of innovation activity (broadly defined) and diverse workers. Starting with the bottom box, a greater agglomeration of heterogeneous people (brain-workers) in a city, together with the institutions and services that support these workers, leads to higher productivity in that city's innovative industries, through the complementarity of heterogeneous brain-workers in innovation activity (*forward linkages*). This, in turn, induces more innovation-based activities and institutions to agglomerate in the city. The resulting increase in innovation activities then creates demand for an even greater variety of brain-workers and supporting institutions in the city, inducing even more brain-workers to agglomerate in the city (*backward linkages*). Through this cycle of agglomeration of heterogeneous brain-workers and innovation activities, the city

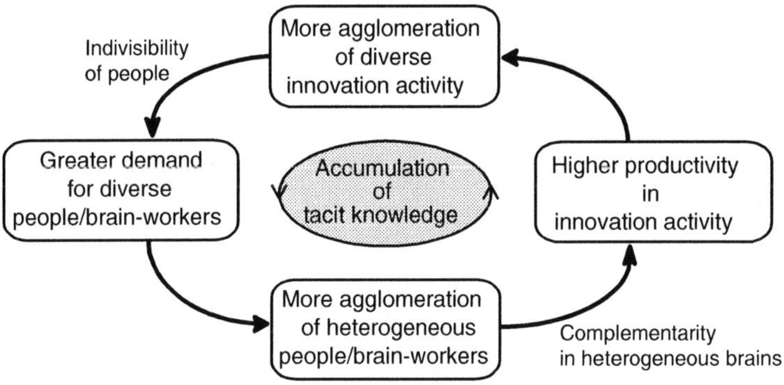

Figure 2.7 Circle of causality in spatial agglomeration of innovation activity and brain-workers
Source: author.

itself accumulates so called *tacit knowledge*, mainly through face-to-face communications among the brain-workers themselves. The local externalities created by the accumulation of tacit knowledge provide the city with a further competitive advantage in innovation activity.

In Figures 2.5–2.7, we have separately considered the diversity of consumer goods, the intermediate goods, and the heterogeneity of people as the basic sources of endogenous agglomeration forces. In reality, of course, the three types of agglomeration forces work together to make industrial clusters and cities grow, while interacting with the rest of the economy.

5. Impact of decreasing transport costs

We now examine the general effects of decreasing transport costs on the spatial distribution of economic activities. As noted in Fujita (2007), the main engine for the current phase of globalization is the steady reduction of *transport costs* (broadly defined). Thus, this aspect of the new economic geography leads to a greater understanding of what has been happening recently in terms of globalization and regional integration.

First, we must recognize that "transport costs" in reality refer to a great variety of diverse costs involved in the movement of goods, services, people, money and capital, as well as information, knowledge and technology. In particular, it is important to distinguish the usual costs for the transportation of goods and persons from the costs related to the transfer of information through various modes of communication. Furthermore, regarding international trade in goods, we must consider, in addition to freight costs, many other forms of trade costs such as tariffs and non-tariff barrier costs, risk from exchange-rate variation, costs for searching and acquiring information necessary for doing business in other countries, and costs arising from different languages and cultures, many of which are difficult to quantify. In addition, before goods produced at a location reach actual consumers, significant costs are usually involved in retail and wholesale distribution.

Over the past half-century, most of these trade costs have been steadily reduced through continuous improvement in transportation and information technologies, together with continuing efforts to lower institutional barriers in international trade, investment and finance.

As shown in Fujita (2007), the reduction of trade costs over the last half-century has resulted in the relative concentration of the world GDP in three regions – East Asia, EU and NAFTA – even as economic interdependency within each region has become stronger. This is almost counter to our intuition that lower transport costs would make geographical accessibility or distance less important, leading to a more even distribution of economic activity in the world. As this example indicates, the impact of decreasing transport costs is in general complex, often involving counter-intuitive phenomena as well as non-monotonic effects. Figure 2.8 helps us to understand why such complex phenomena would occur. This figure shows the general tendencies in the concentration/dispersion of non-land-based activities (land-based = agricultural or resource extraction) that arise from reduction in transport costs.

To make this easier to understand, first let us consider an extreme case in which transports costs are prohibitively high. In this case, since interregional/international trade is essentially impossible, non-land-based activities such as manufacturing and services have no choice but to disperse themselves in proportion to local demand arising from land-based activities and from immobile consumers (due to national borders, for instance), as indicated in the lower right part of Figure 2.8. In this case, since the economy can employ few scale economies or agglomeration economies, the general welfare level of the economy would be relatively low.

Next, suppose that transport costs decrease gradually. At a certain point, the circle of causation in spatial agglomeration of non-land-based activities (represented in Figures 2.5–2.7) goes into effect, leading to the formation of many small cities and industrial agglomerations. As transport costs continue to fall, those activities concentrated in small agglomerations that produce goods or services that are relatively more differentiated (and thus have lower price-elasticities) and/or have relatively lower transport costs will tend to concentrate further into a smaller number of locations. In this way, with the gradual reduction of transport costs (as indicated in the middle part of Figure 2.8), more differentiated goods with low transport costs will initially concentrate into an ever smaller number of cities or industrial agglomerations. Thus, the economy begins to manifest a *hierarchical spatial structure*, in which larger agglomerations provide a greater variety of goods and

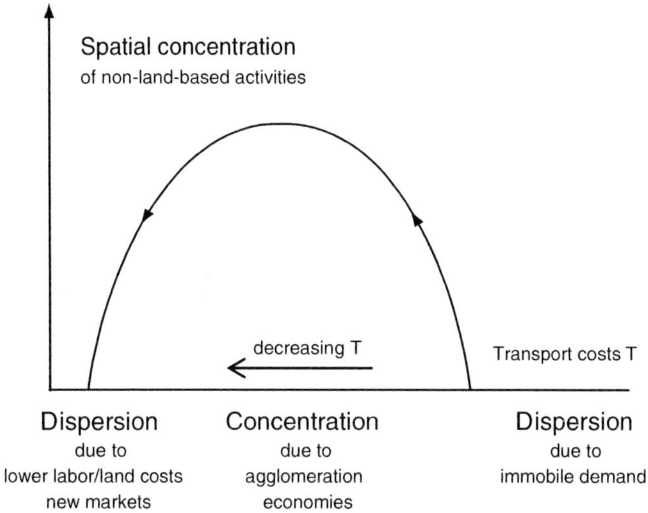

Figure 2.8 Non-monotonic impact of decreasing transport costs
Source: author.

services. With even further reduction in transport costs, we might expect that the economy would come to have a *monopolar spatial structure* in which the most highly differentiated products with relatively low transport costs are all provided from one enormous city or agglomeration.

From the discussion so far, it is important to realize that the signifcant agglomeration of economic activities can emerge only when transport costs become sufficiently low.

With a further reduction in transport costs, however, other factors that we have so far neglected become significant. That is, in a larger agglomeration, land prices naturally become higher, which in turn pushes up wage rates in that agglomeration through higher housing costs and higher prices of non-traded goods and services. Furthermore, the difficulty people have in moving across national borders pushes up labor demand in countries with a larger agglomeration of industries, leading to higher wage rates in these countries. Therefore, as indicated in the left half of Figure 2.8, with a further reduction of transport costs, those activities that use labor (or land) intensively and have relatively low transport costs start moving to smaller agglomerations in peripheral regions/countries. In this way, with a further reduction of transport costs, many industries (or labor-intensive phases of production activities) gradually shift from the core region (or country) first to nearby peripheral regions, then to further peripheral regions. In fact, this is what happened in the "flying geese process" of economic growth in East Asia during the second-half of the last century, as will be further explained in Chapter 4.

As we have seen above, *the impact of decreasing transport costs is non-monotonic*. That is, only with a sufficient reduction in transport costs do agglomeration forces begin to dominate the dispersion force of transport costs, leading the formation of economic concentrations. However, with too much concentration of economic activities in core regions, wage rates there increase together with higher land costs, which tend to push away some of the activities having high labor (or land) intensity to peripheral regions.

Now, a question naturally arises. As we have seen above, when transport costs become sufficiently low, many activities start moving away from large agglomerations (the core) to the periphery. Logically, with the further development of transport and information technologies, will large agglomerations such as major metropolises eventually disappear? That is, will the so called "the death of cities" eventually come to pass? In the very long run, this may indeed happen. Thus far, however, the reality has been the contrary. That is, with the progress of the so called IT revolution over the past three decades, many major metropolises in the world (such as New York, San Francisco, London, Paris, Tokyo, Seoul, Beijing, Shanghai, Hong Kong, and Singapore) are becoming more dominant than before. Furthermore, we have seen in Fujita (2007) that global economic activity (in terms of GDP) has been increasingly converging within and around three regions, i.e., East Asia, EU and NAFTA.

To understand these phenomena in the context of our earlier explanation of the effects of decreasing transport costs, it is important to realize that the recent reduction in transport costs has been accompanied by a broad range of fundamental technological innovations. In particular, the so called IT revolution (more precisely, the revolution in digital technologies made possible by integrated circuits) over the last three decades has contributed not only to the advancement of transport technologies reducing transport costs), but also to a very broad range of innovation in production technology, information processing, communications, and new products. This in turn has led to the creation of a broad range of new economic activities and industries, including the global finance and management, high tech R&D, and software industries. These new activities and industries are generally more knowledge-intensive than traditional ones, and thus tend to agglomerate in major metropolises and advanced regions through the process depicted in Figure 2.7. Therefore, the development of transport and information technologies has thus far tended to increase, not decrease, the dominance of major metropolises and regions involved in knowledge-intensive activities.

Finally, the advancement of transport and information technologies has contributed greatly to the rapid growth of multinational firms, leading to major reorganizations of the global production system and of the division of labor, as discussed in the next section.

6. Growth of multinational firms

Over the last three decades, we have witnessed a rapid increase in the number, size, and capabilities of multinational firms (MNFs). A MNF organizes and performs discrete activities in various countries, which altogether form a supply chain beginning with a product's conceptualization and ending at its delivery. This spatial fragmentation of production aims to take advantage of differences in technologies, factor endowments, or factor prices, as well as market sizes, between countries. Supply chaining is regarded as one of the main ingredients of the process of economic globalization. In particular, MNFs have been playing a dominant role in the economic growth and integration of East Asia. In 2003, for example, 57% of both the exports and the imports of China was conducted by MNFs.

A typical MNF (in a manufacturing industry) keeps its strategic functions such as HQs and R&D in the home country where high-skilled workers are available, while conducting its production activities in "host" countries as well as sometimes in the home country. There exist two primary reasons for conducting production activities in host countries:

Reason 1: to take advantage of low wage rates (and, more generally, low factor prices) in host countries

Reason 2: to have better access to the markets of host countries (and nearby countries)

Typically, when MNFs set up their overseas production plants in developing countries, reason 1 is mainly at work (and reason 2 when production is located in developed countries). However, the two reasons are not mutually exclusive. For example, when a developing country has a large market (e.g. China), the two reasons work together to encourage MNFs to locate production in that country. Another example of the reasons working simultaneously is when Japanese MNFs operate production plants in Mexico or Poland, intending to sell their products in the US or EU market.

Regardless of whether reason 1 or reason 2 is responsible when a MNF sets up production activities overseas, the efficient organization of a MNF becomes possible only when transport costs (for products and intermediate inputs) and communication costs (among fragmented activities) are sufficiently low. Thus, the rapid increase of MNFs in the past few decades owes much to the revolutionary development of communication and information technologies over the past half-century.[6]

7. Conclusion

This chapter has presented a conceptual micro-economic framework for understanding the formation and growth of economic agglomeration and industrial clusters, based on the theory of spatial economics. In particular, the focus of this chapter was on the economic modeling of the endogenous mechanisms that lead to agglomeration. Although only the basic theoretical framework has been presented in this chapter, the articles in the rest of this book present many specific examples of industrial agglomerations in Asia.

Notes

1. Strictly speaking, *the new economic geography* represents a new branch of spatial economics, initiated by Krugman (1991a,b) and others in the early 1990s, which emphasizes the general equilibrium approach to the location of economic activities. For a historical overview of the development of spatial economics in connection with the new economic geography, see Fujita (2005). See also Fujita and Krugman (2003) for a historical review of the development of the new economic geography, and Fujita and Mori (2005) for recent developments in the field.
2. In the remainder of the article, we use the term "industrial agglomeration" instead of "industrial district."
3. For more detailed descriptions of each type of agglomeration, refer to Table 1 in Markusen (1996).
4. See also Fujita and Thisse (2002) and Baldwin et al (2004) for more recent developments in the new economic geography.
5. See Kuchiki (2004) for the discussion of capacity building in the context of Asia's MNF-initiated industrial clusters.
6. See Fujita and Gokan (2005) and Fujita and Thisse (2006) for further analyses of the evolution of MNFs from the viewpoint of spatial economics.

Bibliography

Baldwin, R., R. Forslid, P. Martin, G. Ottaviano and F. Robert-Nicoud. *Economic Geography and Public Policy.* Princeton, New Jersey: Princeton University Press, 2003.

Fujita, M. (ed.) *Regional Integration in East Asia: from the Viewpoint of Spatial Economics.* Macmillan, 2007.

Fujita, M. (ed.) *Spatial Economics* in the series of The International Library of Critical Writings in Economics. Cheltenham, UK: Edward Elgar, 2005.

Fujita, M. and J.-F., Thisse. "Globalization and the Evolution of the Supply Chain: Who Gains and Who Loses?" *International Economic Review* 47(3), 811–836, 2006.

Fujita, M. and J.-F. Thisse. *Economic of Agglomeration: Cities, Industrial Location and Regional Growth.* Cambridge: Cambridge University Press, 2002.

Fujita, M. and P. Krugman. "The New Economic Geography: Past, Present and The Future." *Papers in Regional Science* 83, 139–164, 2003.

Fujita, M., P. Krugman and A. J. Venables. *The Spatial Economy: Cities, Regions and International Trade.* Cambridge, MA: MIT Press, 1999.

Fujita, M. and T. Gokan. On the evolution of the spatial economy with multi-unit multi-plant firms: the impact of IT development, *Portuguese Economic Journal* 4, 129–156, 2005.

Fujita, M. and T. Mori. Frontiers of the New Economic Geography, *Papers in Regional Science* 84 (3), 377–405, 2005.

Fujita, M., V. Henderson, Y. Kanemoto and T. Mori. "Spatial Distribution of Economic Activities in Japan and China" in V. Henderson and J. Thisse (eds), *Handbook of Urban and Regional Economics*, vol. 4, North-Holland, Chapter 6, 2912–2977, 2004.

Hall, R. E. and C. I. Jones. "Why Do Some Countries Produce So Much More Output Per Worker Than Others?" *Quarterly Journal of Economics* 114, 83–116, 1999.

Henderson, J. V. "Medium Size Cities." *Regional Science and Urban Economics* 27, 583–612, 1997.

Krugman, P. "Increasing Returns and Economic Geography." *Journal of Political Economy* 99, 483–499, 1991a.

Krugman P. *Geography and Trade*. Cambridge, MA: MIT Press, 1991b.

Kuchiki, A. "A Flowchart Approach to Asia's Industrial Cluster Policy" in A. Kuchiki and M. Tsuji (eds), *Industrial Clusters in Asia*. Macmillan, Chapter 4, 193–225, 2004.

Markusen, A. "Sticky Places in Slippery Space: a Typology of Industrial Districts." *Economic Geography* 72, 293–313, 1996.

Mori, T., K. Nishikimi and T. E. Smith. *Some Empirical Regularities of Spatial Economies: a Relationship between Industrial Location and City Size*. Discussion Paper No. 551, Institute of Economic Research, Kyoto University, 1984.

Porter, M. E. *On Competition*. Cambridge, MA: Harvard Business School Press, 1998.

Piore, M. and C. Sabel. *The Second Industrial Divide: Possibilities for Prosperity*. New York: Basic Books, 1984.

Saxenian, A. *Regional Advantages: Culture and Competition in Silicon Valley and Route 128*, Cambridge, MA: Harvard University Press, 1994.

Part II

Case Studies

3

Flowchart Approach to Industrial Cluster Policy: Guangzhou's Automobile Industry Cluster

Akifumi Kuchiki and Hiroyuki Tsukada

1. Introduction

Whether a region's economy can survive in the competitive world may depend on the success or failure of industrial cluster policy. There are various types of industrial clusters in Asia – the automobile industry cluster in the suburbs of Bangkok in Thailand, the electronics industry cluster at the Thang Long Industrial Park (TLIP) in Hanoi, Vietnam, the software industry cluster in Beijing, China, and the electronics cluster in Penang, Malaysia. While classic central government industrial policy – a policy of selective intervention – fosters a specific industry over an entire nation, local governments' industrial cluster policies are aimed at activating regional economies. Industrial cluster policy is undertaken not by central governments, but by local ones, partly because economies all over the world have become decentralized as globalization has advanced (Fujita, Krugman, Venables 1999). The critical question for industrial cluster policy, naturally, is whether such policies can actually succeed in encouraging the formation of industrial clusters.

According to Fujita (2003), the central issue in the field of spatial economics is to clarify the mechanisms of industrial agglomeration, while the objective in the study of industrial clustering is to determine how firms in industrial clusters become more innovative as a result of locating in the cluster. Here, an industrial cluster is defined as an industrial agglomeration that induces firms to innovate and to create knowledge and technology in order to be competitive with other firms.

The question of whether national industrial policy is effective as a growth strategy for a nation has been the subject of much discussion. For example, the World Bank published a book titled "East Asian Miracle" in 1993, which tentatively concluded that industrial policy can be effective if central government officials in countries like Korea and Japan are capable of effectively implementing the policy. But Stiglitz and Yusuf (2001) concluded after the Asian Currency Crisis in 1997 that it is not efficient for central governments to selectively allocate resources to specific industries and firms. That book pointed out that industrial clusters are more important than industrial policy for economic development.

Porter (1998) proposed a "diamond approach" for a cluster to achieve competitive advantages over other clusters. The approach explains that, after a cluster is formed, the cluster becomes innovative if four conditions are satisfied. This approach, however, is not a model of how to create a cluster through policy measures. Baldwin (2003) discussed industrial cluster policy not practically but theoretically. The question of whether industrial cluster policy is effective in forming industrial clusters has thus remained unanswered.

As shown in Figure 1.1 in Chapter 1, Kuchiki (2005) proposed a "flowchart approach to industrial cluster policy," theorizing that industrial cluster policy can be effective in forming industrial clusters by establishing export-processing zones, building capacity, and inviting anchor firms. Capacity includes infrastructure, institutions, human resources, and living conditions. Appendix of this book delineated this flowchart approach by proposing sufficient conditions for the formation of the kind of industrial clusters typical of Asian manufacturing industries to enhance regional growth. The typical Asian model of industrial cluster policy included the definition of an industrial zone as a "quasi-public good." It was shown that cluster policy enhances economic growth under a production function that includes the "increasing returns to scale" of an anchor firm, and that the decision-making of the anchor firm's related firms to invest in the clustered region depend on this anchor firm creating a critical mass of "scale economies." Kuchiki (2003) illustrated the successful cases of Canon in Hanoi, Vietnam, and Toyota in Tianjin, China. Canon and Toyota, both Japanese firms, functioned well as anchor firms in the flowchart model.

The purpose of this chapter is to apply the flowchart approach to industrial cluster policy to the case of the automobile industry in Guangzhou in China, and to make it clear that the role of the anchor firms Honda, Nissan and Toyota is key to cluster policy-driven industrial agglomeration in this region. We demonstrate the effects of Honda, Nissan and Toyota on the success of cluster policies.

The leadership of the government of Guangzhou Municipality has played a crucial role in the success of the region's industrial cluster policy. The roles of this local government are (1) management of industrial zones, (2) promotion of joint ventures between state-owned enterprises and foreign investors, (3) promotion of joint ventures between Guangzhou

Automobile and foreign investors such as Honda, Nissan, Toyota, Isuzu and Hyundai, and (4) support for foreign investors. It is not central but local governments that play comprehensive and crucial roles in compensating for market failures, such as coordination failure and provision of public goods, in the implementation of cluster policy. In sum, the role of local government is key in industrial cluster policy, in contrast with classic industrial policy, where the central government plays the key role.

The flowchart approach in this chapter leads to the following two conclusions. First, a careful ordering of policy measures is needed to implement industrial cluster policy. That is, governments must consider the prioritization (i.e. the timing and order) of policy measures if cluster policy is to succeed. Second, the roles of economic agents outside the central government, such as local governments, actors such as NGOs in the quasi-public sector, and private firms, are important for building the capacity required to implement industrial cluster policy.

This chapter is organized as follows. Section 2 explains the basic model of the flowchart approach to industrial cluster policy. Section 3 applies this model to the automobile industry in Guangzhou and illustrates the effects of Honda, Nissan and Toyota as anchor firms in the industrial cluster policy of Guangzhou Municipality in China. Section 3 also examines the technological innovations in the automobile industry cluster in Guangzhou Municipality and the constraints on the city's further development. Section 4 emphasizes the importance of the role of local government in industrial cluster policy. Section 5 concludes the chapter.

2. A general model of the flowchart approach

In the manufacturing industry, an anchor firm is a firm that assembles products composed of many parts and components. An anchor firm in the automobile industry requires more than twenty thousand parts to assemble one car, and an anchor firm in the printer industry uses about eight hundred parts per printer. The related firms who supply these parts and components to the anchor firm will move into an industrial zone where their anchor firm is a tenant, if sufficient conditions are met. An industrial cluster will thus be formed around the industrial zone by the agglomeration of the anchor firm's related firms. This agglomeration will bring regional growth to the cluster. We can observe many examples of this type of industrial clustering, such as the automobile industry clusters in Tianjin and Guangzhou in China. In sum, agglomeration with an anchor firm setting up the conditions for the arrival of related firms will bring regional growth.

As shown in Figure 1.1 in Chapter 1, this section proposes a general model of the flowchart approach to industrial cluster policy, considering industrial cluster policy as a regional growth strategy rather than a national industrial policy. A sufficient condition for the success of industrial cluster policy is the satisfaction of the following conditions in the proper order: (1) industrial zones, (2) capacity building, and (3) anchor firms.

Many industrial clusters in East Asia have satisfied the conditions above. A local government first constructs an industrial zone as a "saucer" to invite foreign investors. Next, the government builds capacity to improve business and living conditions for the foreign investors. The elements of capacity building include (i) constructing physical infrastructure, (ii) building institutions, (iii) developing human resources, and (iv) creating living conditions amenable to foreign investors. Physical infrastructure refers to roads, ports, communications, and so on. Institutional building, also crucial to success in inviting foreign investors, includes streamlining investment procedures through one-stop services, deregulation, and introduction of preferential tax systems. Human resources, which are usually a necessary initial condition for foreign investment, include unskilled labor, skilled labor, managers, researchers, and professionals. Living environment includes, for example, the provision of hospitals and international schools. An anchor firm will be typically ready to invest after all of this capacity building has been carried out.

3. Application of the Flowchart model to the automobile industry in Guangzhou

3.1. Conditions in Guangzhou

3.1.1. Industrial zones

This section outlines the automobile industry cluster in Guangzhou City, using Table 3.1 from this chapter and Figure 3.1 from Chapter 1. The basic model of the flowchart approach begins with the construction of industrial zones. Both the public sector and the private sector can efficiently provide industrial zones as quasi-public goods (see Kuchiki (2005a); this issue is also discussed in Appendix.) The public sector typically provides these in advance of the private sector, since constructing such zones is often not a profitable activity. Guangzhou Municipality constructed several industrial zones for the automobile industry in southern China, as shown in Table 3.2.

3.1.2. Capacity building

Rows 1–7 in Table 3.2 summarize the industrial zones and capacity building of the automobile industry cluster in Guangzhou Municipality. Recall that capacity building is defined as (a) improving physical infrastructure, (b) reforming institutions, (c) developing human resources, and (d) improving living conditions. We clarify the challenges for future capacity building

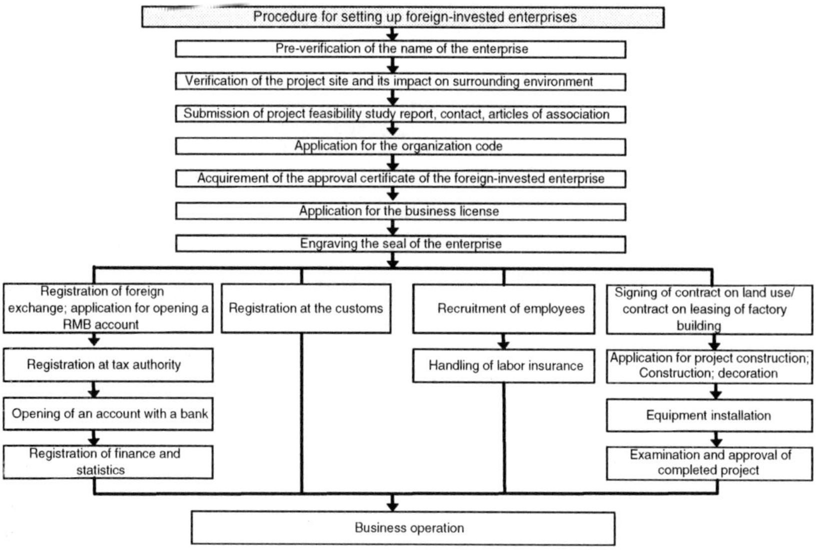

Figure 3.1 One-stop services

in Guangzhou and identify the economic agents responsible for meeting each.

Capacity building is needed to invite an anchor firm and its related firms (including banks and logistics firms). The role of the central government in implementing cluster policy is not as large as its role in implementing industrial policy; institutional reform is the central government's most important job in creating clusters. In contrast, local governments in China play a central role in implementing industrial cluster policy, as we explain in section 4. The leadership of local government heads – mayors in particular – is one key to a cluster policy's success.

Guangzhou Municipality plays important roles in forming its automobile industry cluster. Its challenges for 2005 included the following.

- Infrastructure: addressing the shortage of electricity and construction of highways;
- Institutions: creating a preferential tax system, preferential treatments for commission fees, and one-stop services;
- Human resource development: dealing with a rise in labor costs and the shortage of the Chinese interpreters of the Japanese language;
- Residential conditions: construction of subways.

Table 3.1 Automobile industry cluster in Guangzhou

	Toyota	Honda	Nissan
Joint venture company	Guangzhou Toyota Motor Co., Ltd.	Guangzhou Honda Automobile Co., Ltd.	Dongfeng Nissan Motor company
Establishment	2004 September 2006 May	1998	2004 May
Joint venture partner	Guangzhou Automobile (JVs with First Automotive Works in Tianjin)	Guangzhou Automobile, Dongfeng Automobile (Engines)	Dongfeng Automobile
Main products	Camry	Accord, Fit	Sunny, Teana
Annual production volume	100,000 (2006) Production capacity: 300,000	240,000 (2004). Second plant is under construction (Scheduled to complete in 2006)	150,000
Location of headquarters	Southern coast of Guangzhou city (Nansha district: 797km^2)	Center of Guangzhou city (Guangzhou Economic Development Zone)	North of Guangzhou city (Huadu district: 50 km^2)
Agglomeration of components makers	Hahsha district (12 keiretsu companies), Shunde district, Foshan city (6 keiretsu companies)	Guangzhou Economic Development Zone	Huadu district
Main components makers	Japanese makers	Japanese makers	Taiwanese makers Japanese makers

Notes: Korea Hyundai Motor and Guangzhou Automobile produce commercial vehicles in Huadu. (Total investment: 135 billion yen) Hyundai Motor and Beijing Automobile Investment Co. produce passenger vehicles.

Source: Based on Tsukada (2005), Kuchiki arranged partly.

One-stop services are crucial to a successful industrial cluster policy. Figure 3.1 shows the process of one-stop services in Guangzhou Municipality.

3.2. The automobile industry in Guangzhou

Honda, Nissan and Toyota, Guangzhou's anchor firms, are automobile assemblers. Their first tier related firms had moved into Guangzhou by 2005. Automobile assembler anchor firms have difficulty producing new car models without the cooperation of their first tier related firms.

The automobile cluster in Guangzhou Municipality is in the process of expanding not only to the nearby municipalities of Foshan and Shunde, but

Table 3.2 Industrial zones and capacity building

Zone	**East auto-industry belt**			**South of Guangzhou**		**Northwest of Guangzhou**
Name of districts	Guangzhou economic &technological development district	Xingtang Industrial Processing Zone of Zengcheng City	Huangpu district	Nansha international automobile industry park	Panyu automobile high-tech industry base	Huadu auto city
Infrastructure: (i) Electricity supply	A 110 thousand KVA(1) trsfmr substa.(2) and completed underground electrical network with 2-circuit electricity supply	A 500 thousand KVA trsfmr substa. and 4 trsfmr substa. of 110 thousand KVA.	Electricity network with 6 of 110 KVA trsfmr substa. a 220 KVA trsfmr substa. and Huangpu Power Plant of 1.1 million KW(3)	Several 110 thousand VA trsfmr substa. Zhujiang Powerplant, with cap. of 1.2 million KW and many thermal power sta.(4) with 50–100 thousand KW	Supplied by Shilou 110 KV Trsfmr Substa., and a new 110 KW trsfmr substa. is under const.(5)	15 trsfmr substa. with 110 KV and a 220 thousand KV trsfmr substa. Power supply rate to enterprises reaches 100% with daily supply of 4,760,000 KW/h
(ii) Water supply	Xinhe waterwks (6) with daily cap. (7) of 300,000 tons	4 large waterwks, Dadun, Xizhou, Xinhe and Shapu, providing daily water cap. of 1.2 million tons	Supplied by Guangzhou municipal waterworks	3 main water pipes from Shawan waterwks serving the whole park with daily supply of 250,000 tons	Supplied by Shawan waterworks with daily cap. of 560,000 tons.	3 waterwks with daily supply cap. of 200,000 tons and water quality reaches state's sanitary standards

Continued

Table 3.2 Continued

Zone	East auto-industry belt			South of Guangzhou		Northwest of Guangzhou
(iii) Sewage treatment	A modern wastewater disposal sta. enabling separate treatment for underground water and sewage water	1st stage of 300,000-ton wastewater disposal sta. is under construction	Water and waste drainage systems is under const., which is available with 1st stage of Da Shadi wastewater disposal sta. of daily cap. of 200,000 tons	Guangzhou Nansha city wastewtr disposal sta., with daily handling cap. of the 1st stage of 100,000 tons	The project of drainage system which adapts the rain water type drainage, capable of 160,000 ton/day wastewtr disposal is under way	150,000-ton/day waster water piping system
Human resource: education	Close to Guangzhou zone of high learning and Guangzhou univ.(8) city. Various middle schs.(9), primary schs., kind ergartens and Japanese schs.	Within 30-minute drive to South China univ. of tech.(10), Sun Yet-Tsen univ., Guangdong univ. of foreign studies, Guangdong commercial col.(11)	Close to Guangzhou univ. city where many high educational and research inst. (12) are located, capable of supplying rich human resources to enterprises.	Yingdong middle schools and Nansha central primary Schools, etc.	Located in the southeast of the University City, only 20-mins drive with rich human resource available	Guangzhou Inst. of Automobile of South China Univ. of Tech., Guangdong Trsp. (13) Tech. Col., Guangdong Trsp. Tech. Col., Guangdong Peizheng business sch.

Living conditions: (i) Environment	Sound environment of green hills and clean water, with villa gardens, noble residences and tourist resorts	Ecological residential areas, such as Country Garden Phoenix City, Jinxiu Heung Kong Xinjiang and Xinshijie Residential area	Residential areas including Wanke city, Huangpu, Jinbi millennium with average estate price at 3,500 yuan/M^2	Composed of river and lake with nice natural environment and well-preserved natural ecological zone	Modern residential areas with nice environment, excellent facilities, and convenient trsp. network are available	The Sunshine residential area is surrounded by favorable environment
(ii) Food & entertainment	4-starred hotel (foreign business center), and a gold course	2 five-star hotels and 3 four-star hotels and 4 golf courses	Famous for Nangang seafood and Bicum lichi, a delicious fruit exclusively from Guangdong Province	Tourist attraction with Nansha Puzhou Park, Waterside Str., Nansha Golf Course, Sunflower Park, Wetland Park, Nansha Hotel, Clifford Hotel, etc.	Lotus Hill, one of the eight new scenes of Guangzhou, and golf club, an ideal place for recreation	8-starred hotels, 2 golf courses and shopping and entertainment facilities in the Sunshine Residential area, Automobile city
(iii) Medical services	GETDD hospital (322 beds) and more than 10 medical inst.	General hospitals with advanced tech. and equipments, several medical inst.	More then 10 hospitals including Sun Yet-Tsen Huangpu affiliated hospital	Nansha hospital	2 A-class hospitals, 1 int'l medical insti. and medical inst. at every town	18 hospitals and the world-level Clifford hospital is under const.

Note: (1) KVA=kilovolt amperes (2) trsfmr substa.=transformer substation (3) KW=kilowatt (4) sta.=station (5) const.=construction (6) waterwks=waterworks (7) cap.=capacity (8) Univ.=University (9) sch(s)=school(s) (10) tech.=technology (11) Col.=College (12) inst.=institution (13) trsp.=transportation

Source: Guangzhou Municipality (2005b).

also to the more distant Dongguan, and now covers almost the entirety of southern China's Guangdong Province. Sumitomo Corporation is planning to link its industrial zone in Hanoi in Vietnam to the automobile cluster in Guangzhou Municipality through logistics. Dongguang formed its own clusters in the electronics and textile industries in the past, but has now begun to form an automobile industry cluster.

3.2.1. Honda's effect

Honda began to produce motorcycles in Guangzhou in 1992. The suppliers of parts and components for Honda motorcycles located in Guangzhou are listed in Table 3.3. It is said that Honda's production of motorcycles in Guangzhou is part of the reason that Honda bought a factory located in the city from France's Peugeot.

Guangzhou Honda Automobile was founded on July 1, 1998; Honda Automobile of China was founded on September 8, 2003. Honda constructed its first and second factories in the region in the center of Guangzhou. Firms related to Honda are located in the Guangzhou Economic and Technological Development Zone (GETDZ), the Zhongshan Torch Hi-Tech Industrial Development Zone, Huadu District in Guangzhou City, and in Dongguan City. Suppliers of Honda are mainly located in the Guangzhou Economic and Technological Development Zone (GETDZ) in Zengcheng City whose area is 1741 square kilometers. Zengcheng, Zhongshan, and Foshan thus represent the major agglomerations of Honda's related firms. As Table 3.1 summarizes, Guangzhou Automobile and Honda are under joint management to produce Accord-type and Fit-type cars. It is noted that Dongfeng Automobile and Honda are under joint management to produce engines. Honda's second factory in Guangzhou started to produce Honda Accord cars in 2005.

Guangzhou Municipality played a crucial role in inviting Honda to the city in 1998. This reinforces our assertion that local government's leadership role is crucial to the successful implementation of industrial cluster policy.

Most of the first, second and third tiers of Honda's suppliers moved into Guangzhou from 2001 to 2005, as shown in Table 3.3. There are six firms of Honda's group in Guangzhou: Guangzhou Honda Automobile, Honda Engineering China (engineering), Dongfeng Honda Engine (engines for passenger cars), Wuyang-Honda Motors (motorcycles), Honda Automobile (passenger cars for export), and Honda Motor. Honda's Japanese suppliers have moved into the Zengcheng and Yonghe Economic Zones.

Firms related to Honda (its Japanese suppliers) include Sanoh Industrial (brake tubes), Stanley Electric (lamps), and Mitsuba Corporation (motors). Furthermore, eleven Honda-related firms constructed their factories in Guangzhou from 2001 to 2004: Imasen Electric Industrial (seat adjusters),

Table 3.3 Honda's related companies in Guangzhou

	Company name	Investor	Products	Foundation
1	Guangzhou Showa Autoparts	Showa, Kanematsu	Shock absorbers	1994
2	Guangzhou Stanley Electric	Stanley Electric, Honda Motor (China) Investment	Automotive lightning products	1999
3	Guangzhou Mitsuba Electric	Mitsuba, Mitsuba China (Hong Kong)	Auto electrical systems, two-wheeled vehicle engine support	1999
4	Guangzhou Sanoh Seikan	Sanoh Industrial	Automotive parts (brake tubes, fuel tubes, other tubular parts)	1999
5	Nippon Leakless	GZ Leakless	Automotive gaskets	1999
6	Auto Parts Alliance (China)	Kikuchi, Takao Kinzoku Kogyo	Aluminum structural assemblies	2001
7	Guangzhou Daiyu Seat	Daiyu	Sewn automotive seat components	2001
8	Guangzhou Marujun	Marujun	Automotive stamping parts	2001
9	NHK-Uni Spring (Guangzhou)	NHK Spring	Coil springs, stabilizers	2002
10	Guangzhou NTN-Yulon Drivetrain	NTN, Yulon Group (Taiwan)	Constant-velocity joints (CVJs)	2002
11	Guangzhou Vigo Stanley Electric	Stanley Electric	Automotive lightning products	2002
12	Guangzhou Nanbu Plastics	Nanbu Plastics	Plastic products	2002
13	Guangzhou GKI Car Interior Parts	Itochu, Kotobukiya Fronte	Automotive upholstery	2003
14	NHK Spring Precision (Guangzhou)	NHK Spring	Suspension springs	2003
15	Honda Automobile (China)	Honda Motor	Motorcycles, automobiles	2003
16	Guangzhou Dempu Automotive Parts	N.A. (*1)	Plastic products	2003
17	Guangzhou Shiroki	Shiroki	Doorframes, moldings	2003
18	Guangzhou Ahresty Automobile Parts	Ahresty	Die casting products	2003
19	Guangzhou You-Ri Automotive Parts	Sumitomo Pie & Tube, Sumitomo, Nippon Steel	Electric-resistance Welded pipes	2003
20	Tokai Rubber (Guangzhou)	Tokai Rubber Industries	Anti-vibration rubber parts, automotive hoses	2003

Continued

Table 3.3 Continued

	Company name	Investor	Products	Foundation
21	Honda Engineering China	Honda Engineering, Honda Motor (China) Investment	Mold parts	2004
22	Guangzhou Neive	Neive	Automobile parts	2004
23	Decker System	Decker, Chinese United Faith	Automobile parts	2004
24	Guangzhou Tech Interior Trim Manufacturing	TS Tech	Sewn automotive seat components	2004
25	Teikuro Guangzhou	Teikuro	Surface treatment for mold parts	2004
26	Guangzhou Bridgestone Chemical Products	Bridgestone, Guangzhou Automobile Group	Urethane foam	2004
27	Guangzhou Fuji Tool	Fujiseiko	Carbide tools	2004
28	Toyo Automotive Parts (Guangzhou)	Toyo Tire & Rubber	Anti-vibration rubber parts	2004
29	Ring Techs Automobile Products	Ring Techs, Metal One	Steel wheels for passenger cars	2004
30	Tsuchiya (Guangzhou) Automotive Components	Tsuchiya	Plastic products	2004
31	Akebono Co. (Guangzhou)	Akebono Brake Industry	Brakes	2004
32	Ohashi Technica (Guangzhou)	Ohashi Technica	Automobile parts	2004
33	Guangzhou Uchiyama Manufacturing.	Uchiyama Manufacturing	Gaskets, seals	2004
34	Tocalo & Han Tai	Tocalo, Han Tai	Surface modification	2005
35	Guangzhou Hori Automobile Glass	Hori Glass	Plastic parts for windshields	2005
36	Guangzhou Linjun Automobile Internal Decoration	Hayashi Telempu, Guangzhou Automobile Group	Automotive upholstery	2005
37	Guangzhou Ninomiya Cold Forging Automotive Parts	Ninomiya	Cold forged products	2006
38	Tigerpoly Manufacturing	Tigers Polymer	Rubber seats, plastic hoses	2006

(*1)=not available.

Source: This table is compiled based on Inagaki(2004) and the data obtained from Shunde District, Foshan Municipality in 2005.

Keihin Corporation (engine peripheral devices), Yachiyo, Industry (fuel tanks), F-Tech (break pedals), Hirata Technical Hongo (car body frame components), Kikuchi Takao (car body frame components), Toyo Radiator (radiators), Ogura Clutch (clutches), Nissin Kogyo (break system), Marujun (press parts), and Honda Engineering (engineering). In sum, Honda can procure most of the main components of its cars within the Guangzhou area.

We now explain Honda's effect on industrial cluster policy and industrial agglomeration in Guangzhou, using Figure 3.2. The flow of Honda's effect started from its construction of the first factory in the Guangzhou Economic and Technological Development Zone in 1998 and the second factory in the Zhengcheng Industrial Zone in 2005.

Honda considers Guangzhou City's capacity building, such as facilitation of infrastructure, to be extremely helpful. Problems with capacity often occur. These include shortages of roads, railways, electricity, and other infrastructure, as well as problems with institutional tax reforms, wage rises, and a shortage of Chinese interpreters of the Japanese language. Guangzhou Municipality has solved problems in the past in a timely manner, which is one reason why Honda is increasing the number of its factories in the area in spite of the existence of the aforementioned issues.[1]

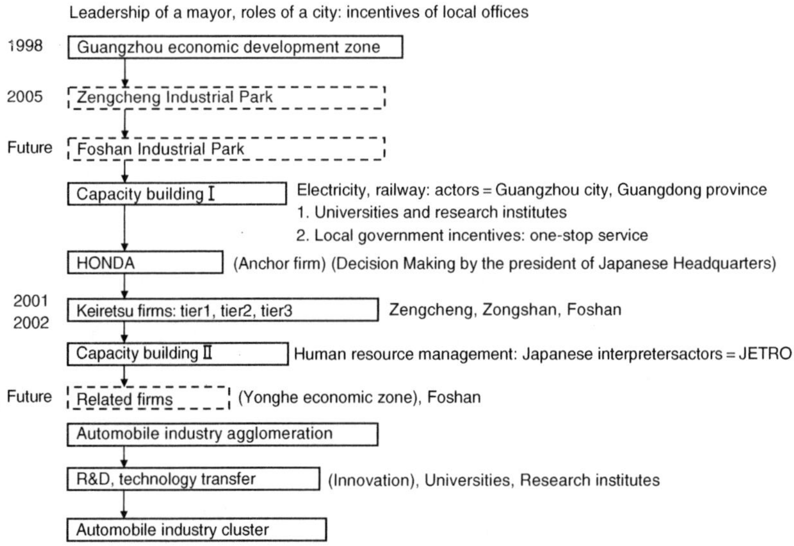

Figure 3.2 Flowchart approach to industrial cluster policy in Guangzhou Honda's effect

Source: author.

The capacity building necessary for Honda's operations involves increasing electricity production and constructing a railway. In 2004, Honda was forced to operate its factory on Saturdays and Sundays and to substitute holidays on weekdays due to a shortage of electricity. Honda, which uses railways to distribute its products, asked Guangzhou Municipality to extend a railway to a location near the factory. Guangzhou Municipality is thus an agent that facilitates infrastructure building, and thus a key player in cluster policy implementation.

Honda moved into Guangzhou by buying a factory from Guangzhou Peugeot, even though Honda realized that the factory did not fit Honda's own factory designs. This purchase is one reason why Guangzhou Municipality reacts favorably to Honda's requests. Honda, for its part, is satisfied with the capacity building in Guangzhou Municipality. The last column of Table 3.3 shows Honda's effect on industrial agglomeration in Guangzhou from 1999 to 2005.

3.2.2. Nissan's effect

Nissan's facilities are located in the Huadu District of Guangzhou. Nissan started to produce cars there in 2004. The area is 50 square kilometers, and a second factory went under construction in 2005. Several firms providing parts and components to Nissan are located in Huadu District.

Dongfeng Automobile and Nissan produce Sunny-type and Tiida-type cars under joint management. Nissan established a joint venture with Dongfeng Automobile in Wuhan, Hubei Province, owning half of the joint venture's 16.7 billion yuan in capital, in June, 2003. Figure 3.3 shows the agglomeration structure of firms related to Dongfeng Nissan Motor Corporation.

The firm, previously named Dongfeng Automobile, produces commercial cars, passenger cars, components and equipment. The firm began operation on May, 2004 and changed its name to Dongfeng Nissan Automobile in

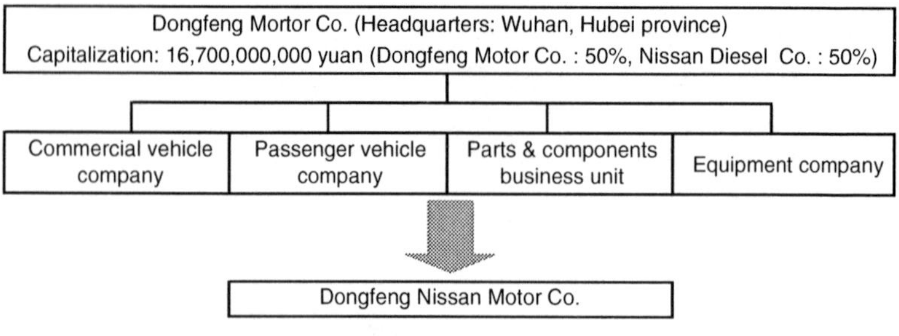

Figure 3.3 Dongfeng Nissan Diesel Motor Co.

Source: Guangzhou Municipality (2005a).

April, 2005. Dongfeng Nissan Automobile operates factories producing vehicles and engines, and an R&D center. Thirty-two of Dongfeng Nissan's parts suppliers are located in southern China, twenty-eight of them in eastern China, and twelve of them in other cities. In 2005, Dongfeng Nissan Automobile purchased 67% of its parts from southern China and 20% from eastern China, in money terms (Guangzhou Automotive Development Forum (2005)). This indicates that Dongfeng Nissan Automobile purchased large quantities of high value-added parts from suppliers in southern China, including Guangzhou City.

According to Inagaki (2004), the following seven firms of Nissan's *keiretsu*,[2] or group firms, moved to Guangzhou city from 2003 to 2005: Hitachi Unisia Automotive (drive train equipment), Kinugawa Rubber Industrial (body seals), Yorozu (suspension parts and modules), Unipress (press forming materials), and Calsonic Kansei (air conditioning systems) in Huadu District.

According to the Guangzhou Automotive Development Forum (2005), Dongfeng Nissan's reasons for investing in Guangzhou's Huadu District were as follows: first, it was effective to set up the firm's new automobile plant by making use of the assets of Aeolus Automobile Co., a former joint venture of Dongfeng Nissan; second, an engine plant and a R&D center, which help develop domestic production capability, should be efficiently located near the main automobile plant; third, both the Guangzhou Municipality and the district government of Huadu were extraordinarily supportive. This third reason was crucial to Nissan's decision to invest in Huadu.

The reasons suppliers of Dongfeng Nissan invested in southern China are as follows: first, the location is close to their customer, Dongfeng Nissan; second, they feel a sense of security due to the presence of so many other Japanese firms; third, the local government supports them by facilitating infrastructure development and offering preferential tax treatment. In sum, Dongfeng Nissan Motor Co., Ltd. and its related firms invested in Huadu District mainly because the local government strongly supported them (Guangzhou Municipality (2005)). The last column of Table 3.4 show's Nissan's foundation of suppliers, which demonstrates Nissan's effect on industrial agglomeration in Guangzhou from 2003 to 2005.

3.2.3. Toyota's effects

The Toyota group in Japan consists of fourteen firms, including Denso, Aichi Steel, Aisin, and Toyota Tsusho. Toyota's supplier group, called Kyohokai, consisted of 208 firms as of 2004, including Aisan, Koito, Tokai Rubber, and Bridgestone. Toyota and Guangzhou Automobile have a joint venture in Guangzhou. As shown in Table 3.1, Toyota is scheduled to produce Camry-type cars in Guangzhou in 2006. Toyota maintains a factory in a 797-square-km industrial zone in Nansha in southern Guangzhou City. The area of Foshan Shunde is 806 square kilometers. Thirteen of Toyota's first

Table 3.4 Nissan's related companies in Guangzhou

	Company name	Investor	Products	Foundation
1	Dongfeng Motor	Nissan Motor, Dongeng Motor	Passenger cars, commercial vehicles	2003
2	Unipress Guangzhou	Unipress	Car body components	2003
3	Yorozu Bao Mit Automotive	Yorozu, Shanghai Bao Steel Int. Economics & Trading, Mitsui & Co.	Suspensions	2003
4	Guangzhou Mitsuike Autoparts	Mitsuike	Metallic dies	2003
5	Dongfeng Passenger Vehicle Research & Development Center	Dongfeng Nissan	Research and development center	2004
6	Guangzhou Hitachi Unisia Automotive Parts	Hitachi Automotive Systems	Power steering system	2004
7	Fuzhou Fukwang Rubber & Plastic	Kinugawa Rubber Industrial	Body sealing products	2004
8	Guangzhou Hua Jing Automotive Steelparts	Mitsui & Co.	Steel plates	2004
9	Kiriu-Lioho	Kiriiu, Lioho Machine Works (Taiwan)	Brake parts	2004
10	Guangzhou Nishikawa Sealing Systems	Nishikawa Rubber	Sealing parts	2004
11	Alpha (Guangzhou) Automotive Parts	Alpha, Marubeni Vehicle	Key sets	2004
12	Xingguang (Guangzhou) Autoparts**	Hoshi Diecast Kogyosho	Die-castings	2004
13	Tacle Guangzhou Automotive Seat	Tachi-S	Automobile seats	2004
14	Guangzhou Kasai Automotive Interior Trim Parts	Kasai Kogyo	Interior parts	2004

Continued

Table 3.4 Continued

	Company name	Investor	Products	Foundation
15	Calsonic Kansei (Shanghai) (Huadu Branch)	Calsonic Kansei	Module products	2005
16	MI Steel Processing Guangzhou	Marubeni Itochu Steel	Steel plates	2005
17	Cast Autoparts**	Daiichi Kinzoku	Automobile parts	2005
18	Fuji Autotech Guangzhou	Shanghai Min Fang Auto Parts, Fuji Kiko, Tachi-S	Automobile seats	2005
19	Calsonic Kansei (Guangzhou) Components	Calsonic Kansei	Instrument panels, exhaust products	2005
20	Guangzhou Mahle Filter System	Mahle Filter Systems Japan	Air filters, motor heat insulation boards	2005
21	Tachi-S Trim Guangzhou	Tachi-S	Automobile seats	2005
22	Engine Factory of Dongfeng Nissan	Dongfeng Nissan	Engines	2006
23	Toyo Quality One Guangzhou**	Toyo Quality One, Nagase & Co.	Automobile devices	2006
24	Guangzhou Nanjo QuanXing Autoparts**	Taiwan QuanXing, Nanjo Kogyo	Automotive seats	2006
25	Guangzhou Kanto Autoparts**	Kanto Auto Works	Auto bodies and parts	2006

**= Author's translation.

Source: This table is compiled based on Inagaki(2004) and the data obtained from Huadu District, Guangzhou Municipality in 2005.

tier keiretsu firms are located in the same district as Toyota in Nansha, and many of Toyota's related firms are located in Shunde and Foshan near Nansha. The second and third tier keiretsu firms now provide their products to the first tier firms of not only Toyota but also Honda and Nissan, a practice that represents the partial collapse of the traditional keiretsu relationships among Japanese firms.

Figure 3.4 of the flowchart approach demonstrates Toyota's effect on the Guangzhou cluster. A highway from Nansha to the center of Guangzhou was constructed in 2005, because it was needed for construction of Toyota's factory – thus, the bridge is clearly an element of capacity building. A subway

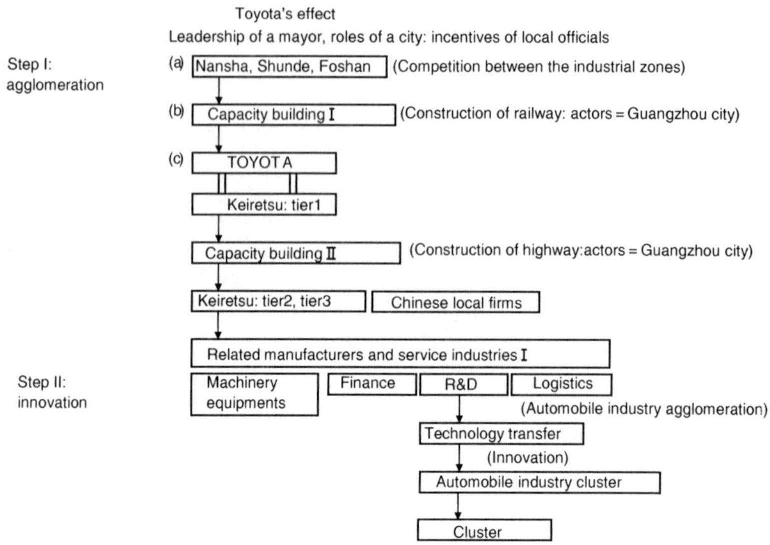

Figure 3.4 Automobile industry cluster
Source: author.

for Toyota's workers to commute is scheduled to be constructed in 2006. Toyota's first tier keiretsu firms are scheduled to construct factories in Nansha, Shunde, and other Toyota areas. The second tier and the third tier firms, together with Chinese firms, are also expected to agglomerate in Guangzhou.

According to a Toyota staff member, Toyota circumspectly requested its keiretsu firms to construct their factories in Guangzhou, and also to satisfy Toyota's criteria for both good quality and reasonable prices. The suppliers are forced to buy parts made by Chinese firms in order to meet these criteria, making Guangzhou an attractive location. The last column of Table 3.5 shows Toyota's supplier foundation, and shows Toyota's effect on industrial agglomeration in Guangzhou from 2002 to 2005. Taken together, Tables 3.3, 3.4 and 3.5 present a broad overview of the effects of Honda, Nissan and Toyota on industrial agglomeration in Guangzhou.

3.3. Technological innovation by transfer through foreign direct investment

Figures 3.2 and 3.4 show how the firms in Guangzhou Municipality create new technologies through innovation, leading to the expansion of the

Table 3.5 Toyota's related companies in Guangzhou

	Company name	Investor	Products	Foundation
1	Foshan Melx Leather*	Melx	Leather products for automobile seats	2002
2	Foshan Nichiwa*	Nichiwa	Bolts/nuts/pressed parts	2003
3	Toyota Gosei (Foshan) Rubber Parts	Toyota Gosei Toyota Tsusho	Body-sealing products for automobiles	2004
4	Aisin Seiki (Foshan) Automotive Parts	Aisin Seiki	Engine parts	2004
5	Aisan (Foshan) Auto Parts	Aisan Industry/ Toyota Tsusho	Throttle bodies/ engine valves	2004
6	Foshan Japan Brake*	Japan Brake Industrial	Brake assemblies	2004
7	Foshan Shunde Yinhe Motorcycle*	Axle	Motorcycles	2004
8	Foshan Tokairika Automotive Parts	Tokai Rika/Toyota Tsusho	Switches/steering locks	2004
9	Koyo Lioho (Foshan) Automotive Parts	Koyo Seiko/Koyo Metaltec/ Toyota Tsusho	Bearings	2004
10	Foshan Yutaka Auto Parts	Yutaka Giken	Torque converters	2004
11	Foshan Summit Nikka Mold & Metal Products	Sumitomo	Mold and metal products	2004
12	Toyota Gosei (Foshan) Auto Parts	Toyota Gosei/ Toyota Tsusho	Interior and exterior parts	2004
13	Toyoda-Koki Automotive (Foshan)	Toyoda Machine Works	Power steering parts	2004
14	Takagi Auto Parts (Foshan)	Takagi Seiko/ Pla-Net	Plastic components/ molds	2004
15	Kobe Wire Products (Foshan)	Kobe Steel/Metal One/Kyodo Shaft/ Sugita Wire	Steel bars and steel wires	2005
16	Elastomix (Foshan)	Elastomix	Carbon master batch	2005

Continued

Table 3.5 Continued

	Company name	Investor	Products	Foundation
17	Guangzhou Parkerizing	Nihon Parkerizing/ Asahi Chiyoda Kogyo/Hamamatsu Netsuhori Kogyo/ Parker Netsuyori Kogyo	Surface treatment for automobile parts	2005
18	Toyota Boshoku Foshan	Toyota Boshoku	Oil filters	2005
19	Fujikura Kasei (Foshan) Paint*	Fujikura Kasei/ Kyokuto Boeki Gaisha	Coating for plastics	2005
20	Foshan Unytite*	Unytite	Bolts/nuts	2005
21	Tokai Spring Mfg. (Foshan)	Tokai Spring Mfg./Nagase & Co., Ltd.	Precision leaf springs press products	2005
22	Yamasei Automotive (Foshan)	Toyoda Machine Works/Yamasei Kogyo/The Yokohama Rubber	Power steering parts	2005
23	Sanyo Seiko (Foshan)*	San-Esu	Mold parts	2006
24	Foshan Dongrong Xuri Electrics*	Toei Kagaku Kogyo/Xuri	Automobile parts	2006
25	Foshan Nitigura	Nihon Glassfiber Industrial	Exhaust manifold covers	2006
26	Foshan Shunde Yazaki Auto Parts	Yazaki	Automotive instruments	2006
27	Hegu (Foshan) Automotive Fuel System*	Nichiwa Sangyo	Automobile parts	2006
28	Daiichi Bussan (Foshan)*	Daiichi-Bussan	Insulators	2006
29	Usui Automotive Parts*	Usui Kokusai Sangyo Gaisha	Brake tubes/power steering tubes	2006
30	Enkei Aluminum Products (Guangdong)*	Enkei	aluminum wheels	2006

Continued

Table 3.5 Continued

	Company name	Investor	Products	Foundation
31	Foshan Shune Lefu Hardware Precise Mold*	Osamu Fukui/ Tsuneo Imai Yosuke Fukui/ Masaru Nikamura	mold parts	2006
32	Azimit Auto Parts (Foshan)*	Azimit	mold parts	2006
33	Sugiyama Kogyo (Foshan) Mold*	Sugiyama Kogyo/ Toyota Tsusho	mold parts	2006

* = author's translation.

Source: This table is compiled based on Inagaki(2004) and the data obtained from Nansha Development District, Guangzhou Municipality in 2005.

cluster. A technology base, embodied in human resources, is needed for this innovation. We thus explain how technology transfer works in Guangzhou.

Technology transfers from the Japanese firms of Honda, Nissan and Toyota to Chinese firms proceed via the three routes. First, some Japanese firms foster Chinese technical experts by sending them to factories in Japan for several months to learn advanced production processes. Second, Chinese firms employ Japanese skilled workers at high salaries after those workers are forced to retire from Japanese companies at the age of 60. Third, technology transfers occur through joint ventures between Japanese and Chinese firms. Examples of these include the joint ventures between Dongfeng Automobile and Honda, between Guangzhou Automobile and Honda, between Dongfeng Automobile and Nissan, between Guangzhou Automobile and Toyota, and a plethora of combined joint ventures between Dongfeng Automobile, Honda, Guangzhou Automobile, Nissan, and Toyota. These joint enterprises serve to standardize technology across the industry and across the region.

3.4. Constraints on the cluster's further development

Figure 3.4 shows how we apply our flowchart approach to industrial cluster policy in Guangzhou. Innovation is the second step in Guangzhou's industrial cluster formation. Japanese firms in the R&D industry, the logistics industry, and the finance industry moved in to drive step II by initiating innovative processes. It is notable that this innovation process did not begin in Guangzhou, but was instigated by foreign firms.

Capacity building in the area of human resources requires that Guangzhou increase its numbers of Chinese interpreters of the Japanese language. These interpreters are crucial in allowing Japanese firms to move into Guangzhou, and is also necessary in order to gather local researchers for the joint R&D projects on which innovation depends.

As shown in Figure 3.2, the industrial automobile cluster in Guangzhou is at the stage where innovation is needed for the further development of the cluster. Japanese firms in the automobile industry first moved into the area to seek cheap labor in the 1990s. Nissan produced Sunny vehicles, but the 2004 sales were not favorable, as Sunny was not popular and was not the newest model. Nissan then started to produce its newest model, named Tiida, and sales of these were favorable in 2005. Japanese firms must change their way of thinking, away from producing old-fashioned models in China by using cheap labor, and toward constructing R&D centers in China, creating new types of technology to fit Chinese consumer tastes, and strengthening the competitiveness of their local subsidiaries in China.

Industries related to the automobile industry have also agglomerated in Guangzhou, as shown in Figure 3.4. These include the electronic parts industry, the machine equipment industry, and automobile-related service industries including finance, logistics, R&D, etc.

4. Industrial cluster policy by the local government of Guangzhou city

This section, summarized in Figure 3.6, shows that Guangzhou's role as a leader in the creation of industrial cluster policy has been crucial to the city's successful development of its automobile industry cluster. The Guangzhou Municipality has positioned the Guangzhou Automobile Industry Group (GAIG) at the center of the cluster, facilitating a joint venture between GAIG and Honda in 1998, a joint venture between Honda and Dongfeng Automobile, and two joint ventures in engines and vehicles between GAIG and Toyota in 2004. Other joint ventures have been facilitated by GAIG; Guangzhou Guangqi Toyota Motor was founded on September 6, 2004, GAIG and Hyundai Motor Company signed to establish a joint venture on June 21, 2005, and GAIG and Isuzu established a joint venture in 2000. The municipality also facilitated the establishment of sixteen joint ventures between GAIG's affiliated firms and the following sixteen firms: Denso, Toyota Bosyoku, Hayashi Telempu, T-ST, Stanley Electric, Honda Engineering, Bridgestone, Toyota Tsusyo Corporation, Daiki Alminium Industry, Chuo Precision Industrial, Parker Corporation, Showa Corporation, Mitsuba, Kanematsu Corporation, Nippon Konpo Unyu Soko, and Honda Express.

As shown in Figure 3.5, Guangzhou's automobile industry cluster includes procurement firms and marketing firms. Procurement firms numbered more than three hundred in 2005, in steel, components, rubber, electric and electronic products, upholstery, and other areas. The tenants of Guangzhou Automobile Industrial Park (GAIP) are joint ventures of GAIG and the firms in Table 3.6.

Using Figure 3.6, we show that Guangzhou Municipality has guided the industrial cluster policy in Guangzhou to success. The municipality's industrial cluster policy measures are as follows:

- Establishment and management of industrial zones as quasi-public goods (see Kuchiki (2005a);
- Support for GAIG and Guangzhou Automobile Group Components;
- Promotion of joint ventures between state-owned enterprises and foreign firms.
- Guangzhou's mayor and the local officials of Guangzhou Municipality may have personal incentives to promote these ventures, as doing this may win them promotion. These incentives are thought to have been effective in engendering the success of the industrial cluster policy.

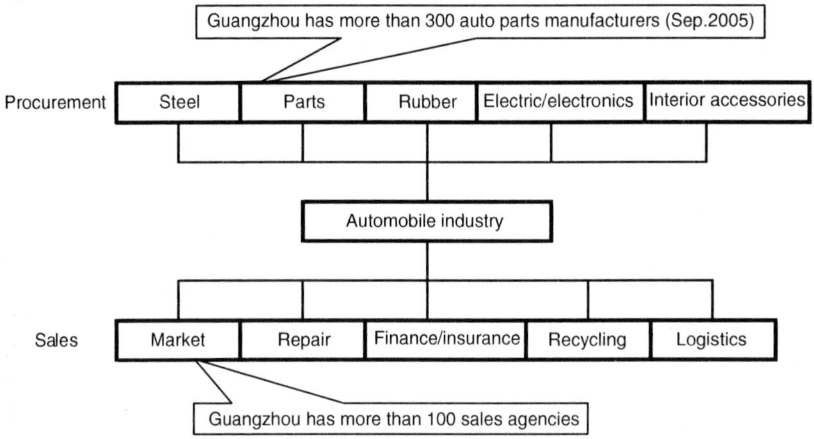

Figure 3.5 Guangzhou automobile industry
Source: Guangzhou municipality (2005).

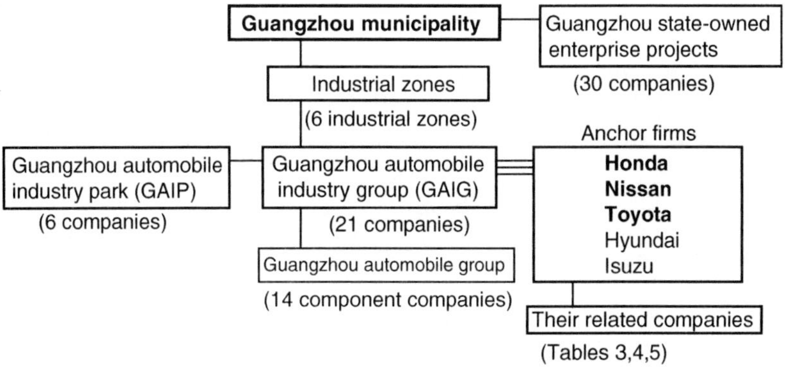

Figure 3.6 Roles of Guangzhou municipality
Source: author.

Table 3.6 Guangzhou automobile industrial park Co. (GAIP)

1	Guangzhou Guangqi Toyotsu Logistics Equipments Co., Ltd.	Deals mainly with design, production, process, sales and after-sales service of industrious jig and fix systems, logistics turnover box and simple transportation equipment.
2	Guangzhou Guangqi Toyotsu Service Co., Ltd.	Provides high quality services to related enterprises, including meal, bus service sanitation, real estate management, etc.
3	Guangzhou Guangqi Toyotsu Resource Management Co., Ltd.	Deals mainly with the recovery, process and sales of auto wastes and ordinary wastes in order to keep the promise of environmental protection.
4	Guangzhou Guangqi Toyotsu Automobile Equipment Co., Ltd.	Deals mainly with the development, design, production sales, installation, maintenance of automobile parts and components and also provides technical service, consultation, after-sales service and lease.
5	Multi-function business Training Center Project.	Provides business training to Guangzhou Toyota project enterprises.
6	Terrestrial Digital Mobile TV Project.	Deal mainly with the development, production, sales and after-sales of terrestrial digital mobile TV terminal.

Source: Guangzhou Automobile Industrial Park Co. (2004).

4.1. Establishment and management of industrial zones for industrial clustering

Guangzhou Municipality manages industrial zones in both Guangzhou and the surrounding cities, as shown in Table 3.2. Guangzhou City has industrial zones in the following six districts:

- Huadu Auto City in Northwest Guangzhou,
- Nansha International Automobile Industry Park, and
- Panyu Automobile High-tech Industry Base,
- Guangzhou Economic and Technological Development District,
- Xiangtang Industrial Processing Zone of Zengcheng City, and
- Huangpu District.

The first three of these are located in South Guangzhou, while the latter three are in the East Auto Industry Belt.

Nissan, its related firms, and Hyundai of South Korea are tenants of Huadu Auto City in Northwest Guangzhou, and Panyu Automobile High-tech Industry Base and Automobile High-tech Industry Base in South Guangzhou. Toyota and its related firms are tenants of Nansha Development Zone. Third, the tenants of the three zones in the East Auto-Industry Belt include the first factory of Guangzhou Honda, the engine factory of Dongfeng Honda, the Honda factory for export, the second factory of Honda, the New Wuyang Honda Motorcycle plant, Baolong Motors, factories of parts manufactures, and marketing firms. It should be noted that the authorities of the industrial zones in Guangzhou City compete each other in inviting foreign firms to their zones.

4.2. Support for the Guangzhou Automobile Industry Group (GAIG) and Guangzhou Automobile Group Component (GAGC)

In 2005, the Bureau of Foreign Trade and Economic Cooperation of Guangzhou Municipality issued a brochure entitled *A Foreign Investment Guide to Guangzhou Auto Industry*, advertising the city's one-stop services (shown in Figure 3.1). These services have helped foreign investors invest in firms related to GAIG.

As shown in Figure 3.7, GAIG, which consists of 21 enterprises in total, is divided into three subgroups: Guangzhou Guangyue Assets Management, Guangzhou Automobile Group, and Guangzhou Motors Group Company. Guangzhou Automobile Industry Group consists of fourteen joint ventures, including those of Guangzhou Automobile with Honda, Showa, and Mitsuba. Guangzhou Guangqi Industrial Development Company has six enterprises in design, food services, recycling, aftercare business, business training, and digital mobile televisions.

AIG has joint ventures with the four anchor firms: Honda, Toyota, Isuzu and Hyundai. Guangzhou Automobile Group Component (GAGC) consists of fourteen firms, including Denso, Stanley, and Bridgestone, shown in Table 3.7. Their products include seats, air-conditioners, lamps, floor carpets, interior components, springs, precise springs, small lamps, urethane, aluminium alloy, interior systems, tire-and-wheel assemblies, and insulators.

4.3. Promotion of establishing joint ventures of state-owned enterprises and foreign firms

Guangzhou Municipality actively promotes the establishment of joint ventures between state-owned enterprises and foreign investors. These foreign investors are given preferential tax treatment and efficient one-stop services. The municipality has established thirty joint ventures

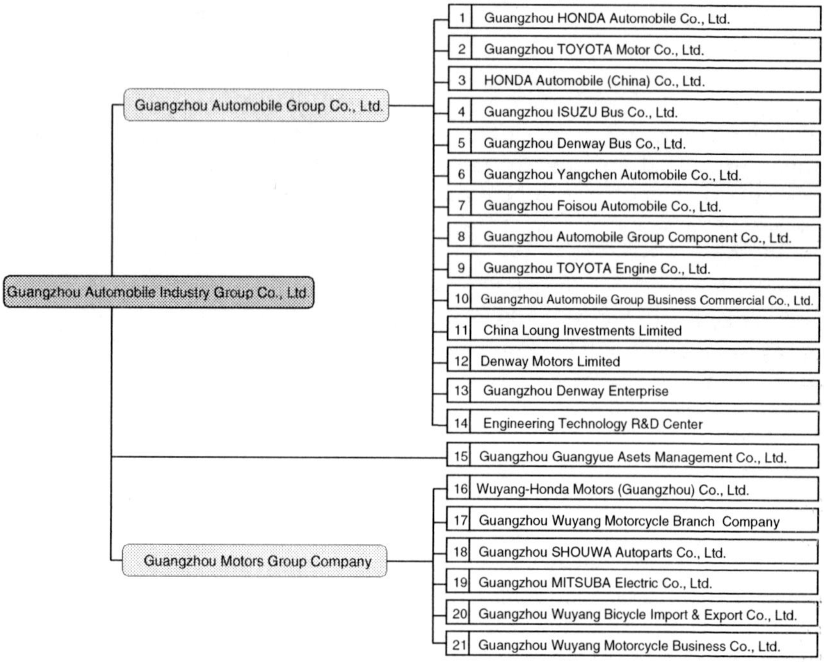

Figure 3.7 Group structure of Guangzhou automobile industry group
Source: Guangzhou Automobile Industry Group Co. (2005c).

Table 3.7 Guangzhou Automobile Group Component Co., Ltd.

1	Guangzhou TS Automobile Interior System Co., Ltd.	Car seats, seat angle modulators
2	Guangzhou Denso Co., Ltd.	Car air conditioners
3	Guangzhou Stanley Electric Co., Ltd.	Lamps
4	Guangzhou Linjun Automobile Internal Decoration Co., Ltd.	Floor carpets
5	Guangzhou Junxing Automobile Components Co., Ltd.	Seat frameworks, car door interior trim panels
6	Guangzhou Jun'an Interior Components Co., Ltd.	Carpets, sound insulation products
7	Guangzhou Huade Automobile Spring Co., Ltd.	Springs
8	Guangzhou Sanda Fine Spring Manufacture Co., Ltd.	Precise springs

Continued

Table 3.7 Continued

9	Guangzhou Sccop Lamps Co., Ltd.	Small lamps
10	Guangzhou Bridgestone Chemical Products Co., Ltd.	Polyurethane
11	Guangzhou Aluminum Smelting Technology (GAST)	Secondary aluminum ingots
12	Guangzhou Intex Auto Parts Co., Ltd.	Seats, door trims and molded headliners
13	Guangzhou Zhongjing Automobile Parts Co., Ltd.	Aluminum wheels
14	Guangzhou Parker Auto Parts Co., Ltd.	Insulators

Source: Guangzhou Automobile Group Component Co. (2005d).

between state-owned enterprises and foreign investors, as shown in Table 3.8.

Many mayors in China are keen to invite foreign investors to their cities. Guangzhou Municipality held an investment seminar in Tokyo, Japan in October 2005, headed by the mayor, Zhang Guangningin. At the seminar, many Japanese investors signed contracts to invest in Guangzhou. It is said that Guangzhou City is trying to become equal in status to Beijing, Tianjin, and Shanghai (Weekly Toyo Keizai, Jan. 6, 2006, p. 35), and that Mayor Zhang Guangning, called the "automobile mayor," is making efforts to invite foreign investors to Guangzhou by complementing the system that invites firms to the province. The position of mayor in China is often a stepping stone to the position of provincial head in China, and thus Mayor Zhang has the opportunity to become head of Guangdong Province by successfully stimulating Guangzhou's economy. His desire for promotion could be one factor in the success of his city's industrial cluster policy. Mayor Zhang targets his policies effectively and exercises strong leadership in part to realize these personal goals. Leadership, and the incentives for leadership, are thus a key to the success of industrial cluster policy.

5. Conclusions

The flowchart approach to industrial cluster policy offers an action plan for appropriately arranging and prioritizing policy measures in a time-ordered series. This chapter proposed a general model of the flowchart approach to industrial cluster policy and applied this model to the specific case of Guangzhou's automobile industry cluster. Japanese firms Honda, Nissan, and Toyota have spurred industrial agglomeration in Guangzhou by locating

Table 3.8 Guangzhou State-Owned Enterprise Project

1	Guangzhou Zhujiang Steel Co., Ltd.	Steel
2	Guangzhou Pearl River Copper Co., Ltd.	Copper
3	Guangzhou Aluminum Materials Plant Co., Ltd.	Aluminum
4	Guangzhou Copper Materials Co., Ltd.	Copper
5	Guangzhou Machine Tools Factory Co., Ltd.	Machine tools
6	Guangzhou Diesel Engine Factory	Diesel engines
7	Guangzhou Cable Factory	Cables, wires
8	Guangzhou Electric Machine Works	Electric machines
9	Guangzhou Jiateli Micro-motor Co., Ltd.	Micro motors
10	Guangzhou Wanbao Refrigerator Co., Ltd.	Refrigerators
11	Guangzhou Wanbao Enameled Wire Co., Ltd.	Enameled wire
12	Guangzhou Tiger Head Battery Group Co., Ltd.	Batteries
13	Guangzhou Eagle Coin Enterprises Group Corp.	Canned foods
14	Guangzhou Baijua Flavours and Fragrances Company Ltd. (G.B F.F.)	Fragrances
15	Guangzhou Watch Factory	Watches
16	Guangzhou Flashlight Industrial Co.	Flashlights
17	Guangzhou South Alkali Manufacturing Co., Ltd.	Sodium carbonate
18	Guangzhou Pearl River Chemical Industry Co., Ltd.	Paint, ink
19	Guangzhou Huangpu Chemical Factory	Chemicals
20	Guangzhou South China Rubber Tire Co., Ltd.	Rubber, tires
21	Guangzhou Building Materials Enterprise Group Co., Ltd.	Building materials
22	Guangzhou Guangxiang Tire Enterprises Group Co., Ltd.	Tires, rubber
23	Guangzhou Baiyunshan Pharmaceutical Co., Ltd./ Guangzhou Baiyunshan Chemistry Pharmaceutical Factory	Medical devices
24	Guangzhou South China Medical Instrument Co., Ltd.	Pharmaceutical
25	Guangzhou Hanfang Modern Natural Medicine Research and Development Co., Ltd.	Pharmaceutical
26	Guangzhou Baidi Biotechnology Co., Ltd.	Bio-pharmaceutical
27	Guangzhou Pearl River Piano Group Co., Ltd.	Pianos
28	Guangzhou Subsidiary Foodstuff Co., Ltd.	Seasonings
29	Guangzhou Foodstuff Group	Edible oil
30	Guangzhou Mr. Sausage Food Co., Ltd.	Meat packers

Source: Guangzhou Municipality 2004.

there, and Chinese firm Guangzhou Automobile, Japanese firm Isuzu, and Korean firm Hyundai have contributed to the effect. The last column of Tables 3.3, 3.4 and 3.5 display the effects of Honda, Nissan, and Toyota on industrial agglomeration in Guangzhou from 2002 to 2005. Guangzhou Municipality implements industrial cluster policy by promoting joint ventures between Guangzhou Automobile and foreign firms.

We obtained the following three conclusions. First, we found significant effects of Honda, Nissan, and Toyota on agglomeration in Guangzhou's automobile industry cluster. Second, we confirmed that our hypothesis of the applicability of the flowchart approach to industrial cluster policy held in the case of the Guangzhou's automobile industry cluster. Third, we established (as shown in Figure 3.6) that local governments play a crucial role in successful industrial cluster policy, and that the head(s) of a local government should be offered incentives, such as promotion, in order to foster industrial clustering. One of the reasons why Guangzhou's automobile industry cluster policy is successful is that the mayor targets his policies and exercises effective leadership, crucial to the success of the city's industrial cluster policy.

Notes

1. We interviewed a staff member of Guangzhou Honda, on 23 August, 2005.
2. A Japanese term describing a loose conglomeration of firms sharing one or more common characteristics. The firms do not necessarily own equity in each other (Investpedia.com).

   Keiretsu (the kanji literally means "series" or "related sequence") is a Japanese term for a set of firms with interlocking business relationships and shareholdings. In Japanese the term also refers to a firm that has many branches (en.wikipedia.org).

References

Baldwin, R. *Economic Geography and Public Policy*. Princeton, NJ.: Princeton University Press, 2003.

Fujita, M. "Significance and Issues of Japanese Industrial Cluster Policy from the Point of Spatial Economics." In A. Yamazaki (ed.), *Japan's Industrial Cluster Strategy*, Tokyo: Yuhikaku, 2003.

Fujita, M., P. Krugman and A. J. Venables. *The Spatial Economy: Cities, Regions, and International Trade*. Cambridge, Mass.: MIT Press, 1999.

Guangzhou Automobile Industrial Park Co. *About GAIP*. 2004

Guangzhou Municipality. *Introduction of Guangzhou State-Owned Enterprise Project*. 2004.

Guangzhou Municipality. *Guangzhou Automotive Development Forum*, Tokyo, September 16, 2005a.

Guangzhou Municipality. *A Foreign Investment Guide to Guanzhou Auto Industry*. 2005b.

Guangzhou Automobile Industry Group Co. *Guangzhou Automobile*. 2005c.

Guangzhou Automobile Group Component Co. *China's Automobile Component Project*. 2005d.

Inagaki Kiyoshi and The 21st China Research Institute. *List of Japanese Invested Enterprises in China* (in Japanese; *Chugoku Shinshutsu Kigyo Ichiran*). Tokyo: Sososha, 2004.

Kuchiki, A. "Agglomeration of Exporting Firms in Industrial Zones in Northern Vietnam." In M. Tsuji, E. Giovannetti and M. Kagami (eds), *Industrial Agglomeration and New Technologies*. Cheltenham: Edaward Elgar, 2007: 97–138.

Kuchiki, A. "A Flowchart Approach." In A. Kuchiki and M. Tsuji (eds), *Industrial Clusters in Asia*. London: Palgrave Macmillan, 2005: 169–199.

Porter, M. E. *The Competitive Advantage of Nations*. New York: Free Press, 1998.

Stiglitz, J. E. and S. Yusuf (eds), *Rethinking the East Asian Miracle*. Oxford: Oxford University Press, 2001.

Tsukada, H. "Change in Southern China." *NEXI*. Nippon Export and Investment Insurance, February, 2005.

World Bank. "The East Asian Miracle: Economic Growth and Public Policy." In *World Bank Policy Research Reports*. New York: Oxford University Press, 1993.

4
Industrial Clusters in the Austin Area: The Austin Technopolis Case Study

Jobaid Kabir

1. Introduction

1.1. Background

As Alfred Marshall observed over 100 years ago, one reason that it was advantageous for similar businesses to cluster in the same city was that workers got together in pubs after work and exchanged ideas. A similar phenomenon seems to have occurred in Austin, Texas, where the likely topic of conversation was software.[1]

In Japan, Malaysia, Singapore, Bangalore, Silicon Valley in California, and in the "Silicon Hills" of Austin, Texas, multiple factors interact to create a regional industrial cluster. Industrial clusters are defined here as geographic concentrations of interconnected companies, including specialized suppliers, service providers, and associated institutions in a particular field.[2] There are certain well-known success stories regarding the development of information technology industry clusters, including those in Silicon Valley and Austin. Agglomeration and clustering theory suggest that an industrial cluster is effective in generating external economies and reducing transportation costs.[3] This study seeks to examine the conditions required for the creation of a cluster, with particular attention to government policies, key individuals, infrastructure capacity, anchor firms, research institutions, venture capital investment, business incubators, and related factors such as perceptions regarding regional quality of life and human resources.

A study of the Austin technology cluster, conducted by the Innovation Creativity and Capital (IC^2) Institute at The University of Texas at Austin (UT) in the late 1980s, developed a conceptual framework for studying the dynamics of high-technology development and economic growth in Austin.

The report referred to Austin as a "technopolis," in which "techno" reflected the emphasis on technology and "polis" referred to the Greek word for city-state, in order to emphasize the connection between the public and private sectors. Austin was successful in its development of the "technopolis wheel," which contains seven segments: the university, large corporations, emerging companies, three levels of government, and various support groups.

The IC^2 study assessed the role and impact of each segment on the emerging Austin technopolis, and underscored the importance of interaction among the different segments. To facilitate interaction, key individuals called "influencers" helped develop new institutional relationships among the segments of the technopolis wheel.[4] This study emphasizes that influencers not only provided leadership within a particular segment, but also networked with other segments and effectively promoted public and private sector entities to jointly develop economic and technology policies. Key findings of the study included the "pivotal role of the research university, the need for continuity in governmental policies, the catalytic role of large technology companies, and the importance of indigenous company development and the need for consensus for the sustainable development of the technopolis."[5] The study summed up three themes in technopolis development that would be relevant to other cases, both within the United States, and worldwide:

- highly coordinated approaches to development,
- the presence of a quality research institution, and
- a network of influencers.[6]

An alternative way to model industrial cluster development is the flowchart approach, developed by Dr. Akifumi Kuchiki of IDE-JETRO of Japan. Kuchiki's flowchart, shown in Figure 4.1, uses four key factors to model the development of an industrial cluster: (a) market and industrial zone(s), (b) capacity building, (c) anchor firms, and (d) related firms.[7] This paper seeks to apply Kuchiki's framework for further analysis of the Austin technology cluster case.

In Kuchiki's model, the special roles of governments and multinational firms are key, particularly the government's decision to designate and build industrial zones or export processing zones, a decision that in theory is based upon market factors for different types of industry. Building an industrial zone leads to expanded infrastructure capacity; once appropriately skilled human resources have been drawn to the area, there is a greater likelihood of attracting an "anchor firm" to the industrial zone. After an anchor firm arrives, it attracts related firms, such as suppliers, eventually gathering enough firms to make the city into an industrial cluster.

As used in the literature of the Austin case study, the term "technopolis" implies interaction among the government, public, and private sectors to

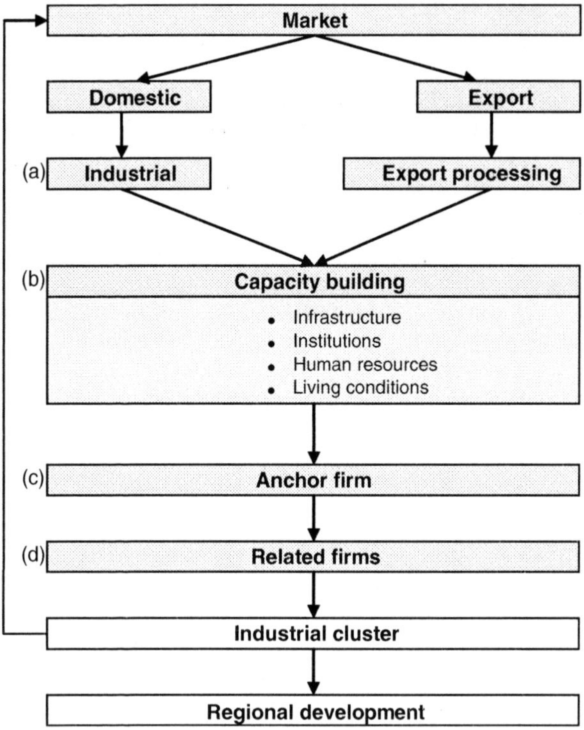

Figure 4.1 Kuchiki's flowchart

Source: Akifumi Kuchiki, Sun Jian and Juan J. Palacios. *Comparison of Industrial Agglomerations Between Asia and the Other Regions*. Japan: Institute of Developing Economies (2005).

spur growth and economic diversification.[8] The term "industrial cluster" in Kuchiki's model is used to refer to an industrial zone that successfully attracts an anchor firm and related supporting firms. The former term implies a level of horizontal interaction and cooperation, while the latter implies more vertical integration and interaction. Nevertheless, we here use the terms technopolis and technology cluster interchangeably, because both describe a level of economic development and growth based on a single type of industry. That industry in Austin is high-technology. For our purposes, the term technopolis can be defined as a city that places emphasis on technology for economic development and growth.

1.2. Objective

The purpose of this research is to examine the development of the Austin cluster, paying particular attention to the most important factors. Our main objective is to use the framework established by Dr. Akifumi Kuchiki of IDE-JETRO of Japan to analyze the technology cluster in Austin, in order to test the robustness of the framework. In addition, this research examines the linkages within and between industrial clusters, which are used to overcome the clusters' deficiencies and augment their strengths. Finally, our conclusion will use the "flowchart model" to demonstrate the dynamism of the industrial cluster, and will identify specific policy priorities for supporting cluster development. We will compare and contrast Kuchiki's model to previous models developed to explain the Austin example. This paper is the first step toward developing new insights into Austin cluster dynamics, into how cluster development changes over time, and into what lessons from the Austin experience can be applied elsewhere.

1.3. History of Austin: pre-existing conditions for a high-tech cluster

This study revealed that the following factors appeared to have impacted on the development of Austin's high-tech industrial cluster:

- Government policies
- UT's role as a research institution
- Austin's quality of life and cost of living
- Infrastructure
- Anchor firms
- Partnerships
- Community organizations
- Key visionaries (influencers)
- Influence of the Silicon Valley model from the 1950s[9]
- Preliminary Conditions

Austin is the capital of Texas, and this status has a major impact on Austin's economy. In 1883 the University of Texas at Austin (UT) was founded, giving Austin its second largest economic engine. For nearly a hundred years, Austin's economy revolved around local services, government offices, and UT. Austin is known for live music, scenic hills, and lakes, and for having a young, active, socially involved population. Prior to the technology boom, local residents had successfully protested to keep out "large smokestack industries" in order to preserve the quality of life in and around Austin, paying special attention to its lakes, parks and walking trails.[10]

Austin has a long history of technology entrepreneurship. In 1955, a UT researcher and professor named Frank McBee, in partnership with others, founded Tracor, Inc., a defense-related research and development and manufacturing company. Tracor became Austin's first high-tech success story, and its only homegrown Fortune 500 Company, until Dell, Inc. earned that honor in the 1980s. At its peak in the late 1970s and early 1980s, Tracor had more than 2,000 employees. More than 25 Austin high-tech companies were spun off Tracor's research and development work.[11] For this reason, Tracor is considered one of Austin's anchor firms.

In 1957, the Austin Chamber of Commerce recommended that the city take steps to attract light manufacturing industries, which laid the foundation for Austin's involvement in the electronics industry.[12] As a result of the Chamber's efforts, IBM opened a manufacturing plant for Selectric typewriters in Austin in 1966. Motorola opened a plant in Austin in 1974, and Advanced Micro Devices (AMD) did the same in 1979. The Chamber of Commerce played a major role in courting Motorola to move to Austin. By the early 1980s, Motorola had established two manufacturing sites and four design centers in Austin.

During the early 1980s, Texans were preparing to celebrate the state's approaching 150th anniversary in 1986 and the centennial celebrations in 1983 of the state's two flagship universities – UT and Texas A&M. A few state and local leaders had the vision to see that, for economic reasons, UT, the city of Austin, and the state of Texas needed to attract new technology and research businesses. Under their leadership, organizations began taking steps to make the state attractive to high-technology companies and make Austin competitive with other states and cities.

At UT, there was a new emphasis on research, which led to the creation of the UT Endowed Centennial Program for chairs, professorships and fellowships in science and technology research. As a result, UT became the nucleus for the development of Austin's technopolis. It played a key role in fostering research and development activities, attracting key scholars and talented graduate students, spinning off new companies attracting major technology-based firms, drawing federal and private sector funding, and serving as a general source of ideas, employees and consultants for high-technology companies.[13]

Between 1984 and 1987, following a plunge in oil and beef prices, Texas experienced an economic recession and budget deficits. Despite the recession, in 1987 "the state began to reverse its policies by increasing funding for higher education and other research support, such as the Advanced Technology and Research Program (ATRP) for attracting the best researchers and students to Texas and expanding the state's existing technology base."[14]

Dell, Inc. is Austin's most famous and visible success story, growing from UT student Michael Dell's garage-based "PC's to order" business to a multibillion-dollar company with $50 billion in revenue, and Austin's largest private sector employer, in the 11 years since its founding in 1984. Dell, which has seen its worldwide market share increase six-fold in the last ten years, serves as a high-tech anchor firm; as such, it is able to support and attract additional businesses to the Austin area.

2. Capacity building

This section describes the various capacity building components of Austin's technology cluster, as defined by Kuchiki's flowchart model.

2.1. Infrastructure

In the last 25 years, Austin has transformed from a small, sleepy college town into a thriving metropolis. The most dramatic increase was from 1980 to 2004, when the population of the Austin Greater Metropolitan Area more than doubled, from 585,000 people to over 1.2 million. This tremendous growth is expected to continue over the next 30 years, during which the population of Austin may reach nearly 3.3 million.[15]

Austin has struggled to keep up with traffic congestion, widening roads, building additional highways, and building a toll way loop around the city. As well as this, in an attempt to ameliorate traffic problems, Austin created a public bus transportation system. Another drag on the city's growth was that Austin's Robert Mueller Municipal Airport was "too small for Austin's dreams."[16] In 1991, when the nearby Bergstrom Air Force Base was listed on the Department of Defense's closure list, the City of Austin officials saw an opportunity to address the airport problem, and almost immediately began making plans to move Austin's airport to Bergstrom. Austin-Bergstrom International Airport (ABIA) began construction in 1993 and was completed in 1999.

Austin's electric utility, Austin Energy, is a community-owned electric utility and a department of the City of Austin. Austin Energy owns 2,626 MW of total generation and operates three natural gas fired power plants. In addition, Austin Energy owns system capacities in the form of one coal fired power plant and one nuclear fuel plant. The Austin Energy electric system serves over 1,000 square kilometers in Travis and Williamson counties.[17]

The Lower Colorado River Authority (LCRA) supplies water to the region, and recreational lake opportunities to its residents, by maintaining the Highland Lakes. The LCRA supplies wholesale electricity to retail utilities and manages water supplies for cities, farmers and industries along the 1,000-kilometer stretch of the Texas Colorado River starting north of Austin and extending to the Gulf of Mexico. It operates six hydroelectric dams on the Colorado River that form the scenic Highland Lakes: the Buchanan, Inks, LBJ, Marble Falls, Travis and Lake Austin dams. The LCRA regulates water discharges to manage floods, and releases water for sale to municipal, agricultural, and industrial users. It also provides electricity to almost 77,000 square kilometers around Austin. Additionally, it operates more than 6,500 hectares of parks and recreational areas around the lakes and river, with more than 40 parks, natural science centers and nature preserves in all.[18]

Austin provides a rich communications infrastructure, with more 200,000 kilometers of fiber network spanning the hills of Austin. Technologies such as Metropolitan Optical Networks (MON), Ethernet over SONET, Optical Ethernet Metropolitan Area Networks, Network Based Virtual Private Networks, VoIP, Dedicated Internet Access over Ethernet, Wireless Fidelity 802.11 and Fiber Channel Metropolitan Area Networks all had their origins in Austin.[19] Furthermore, Austin is the "perfect test market for new wireless technologies," because of its relatively small size, good infrastructure, high income of residents and high-tech-oriented culture.[20]

2.2. Institutions

There were many factors that influenced the development of the Austin technology cluster. Due to the high level of cooperation and coordinated efforts between different institutions, it is difficult to separate and prioritize the contributions of each. However, there are two essential factors that are worth highlighting. First, the university provided the core competence for attracting and developing "knowledge based industry" in Austin. Second, cooperation among local government and support groups was critical to Austin's development.

2.2.1. The university of Texas at Austin

UT was founded in 1883. For most of its first 75 years of existence it had a reputation as a white-collar liberal arts school. The College of Engineering was a relatively small part of the university and did not have a graduate division until the 1950s. In 1958, in order to commemorate UT's 75th anniversary, a "Committee of 75" was set up to review the university's role in the state and its contribution to Texas' economy. The committee recommended that UT transform itself into a research university. The faculty within the College of Engineering, specifically from the Electrical Engineering Department (EE, later renamed Electrical and Computer Engineering, or ECE) played a central role in UT's transformation into a well-respected high-tech research institution.

The EE Department recruited a well-known senior faculty member from MIT, Herbert Woodson. As Chairman of the EE Department, Woodson recruited Ben Streetman, one of the semiconductor industry's leading academic figures and author of internationally used senior-level textbooks. Streetman joined the UT faculty in 1982, and in 1983 he gave a presentation to the MCC Search Committee that demonstrated UT's commitment to microelectronics and semiconductor research.

In 1983, to commemorate UT's centennial, the university established a "Centennial Matching Program" that pledged to match outside contributions toward endowments with income from its own endowment. In 1984 a philanthropist proposed to donate $8 million to any Texas university that could provide plans showing how the money would be spent on research that would propel Texas into the twenty-first century. With the matching grants program, there was a $32 million pool available for the establishment of endowed chairs. UT endowed 16 one million dollar chairs in the College of Engineering with four each for Microelectronics, Computer Engineering, Material Sciences, and Manufacturing, and 16 one million dollar chairs in the College of Natural Sciences with four each in Mathematics, Physics, Computer Sciences, and Molecular Biology.[21]

In 1964 the Department of Computer Sciences (CS) was founded at UT, and by the 1970s, the CS department had established itself as a highly regarded research institution. The department was founded in order to bring UT's diverse computing-related activities together under a single roof. In the 1960s, computing activities and related research work at UT had occurred in widely differing areas under the guidance of faculty members who each had an interest in the application of computer sciences in their core areas of research. Bringing together these activities under the umbrella of a centralized department allowed for cross-fertilization of ideas and creation of an environment conducive to cutting-edge research.

UT has benefited tremendously from the state's Permanent University Fund (PUF), which serves all public universities in Texas, and which had a 2004 book value of $8.1 billion.[22] This public endowment has been crucial to the development of the teaching and research excellence at UT's laboratories. The PUF alone, however, has proved to be insufficient in providing the resources necessary for the development of a world-class university.[23] In 1992–1993 UT's portion of the PUF was $84 million, dropping to $75 million in 1995–1996.[24] In 2004 the entire UT System, including nine universities and six health institutions, received only $341 million for the fiscal year ending August 31, 2005.[25]

UT President Peter T. Flawn (1979–1985) strongly believed in UT's academic mission, which he believed to be at odds with profit-making activities. Though Flawn later joined the effort to bring MCC to Austin, he had initially objected to the idea on the grounds that it was less about furthering UT's academic mission than it was about making profit.[26]

His successor, UT President William Cunningham (1985–1992), began to change the institutional perspective at UT. Under his leadership, UT changed its policies to allow academics to hold equity in the products created through their research, and encouraged them to file for patents for commercially viable technologies. This was in part the result of federal legislation, namely the Bayh-Dole Act, which allowed universities and small businesses to obtain exclusive licensing of new inventions made under federal funding after they committed to diligently developing and transferring the invention to the market for the public good. This helped to foster university-industry research collaborations.[27]

2.2.2. The chamber of commerce

The Chamber of Commerce, after a period of brisk activity from the late 1960s through the 1970s, had fallen into a "vision rut" by the early 1980s, according to C. Lee Cooke, President of the Chamber of Commerce from 1984 to 1987 and later Mayor of Austin from 1988 to 1991.[28] The Chamber's strategic economic plan from the 1950s succeeded in attracting light manufacturing to Austin, but had outlived its purpose.

In 1984, Lee Cooke was a manager at the Texas Instruments plant in Austin. He was also a member of Austin's City Council and worked extensively with the Chamber in its efforts to recruit companies such as AMD to Austin (in 1979) and to influence Motorola's plans to open a new facility in Austin in 1981. When Cooke became president of the Chamber, one of his first tasks was to reinstate economic planning and analysis as a regular function of the Chamber. He hired economists and a consulting firm to conduct a strategic review of Austin and to develop a 15-year plan for economic development. The plan advocated collaboration between public, private and university institutions to create a climate of innovation and entrepreneurialism, as well as programs to recruit and grow technology companies.[29]

Kirk Watson, current Chairman of the Greater Austin Chamber of Commerce, was one of the visionaries in the late 1990s who provided the leadership necessary for the Austin technology industry to grow and gain national and international recognition. During the later 1990s he was able to unite business, community, and environmental leaders in helping Austin's technology industry grow and earn recognition.

2.2.3. The innovation creativity and capital institute

In 1977, Dr. George Kozmetsky founded the Innovation Creativity and Capital Institute (IC^2), a UT-affiliated research institute that was created to facilitate partnerships between academia and industry. Kozmetsky established the institute just before he retired as Dean of the College of Business Administration and the Graduate School of Business at UT, and he provided much of the funding for IC^2 from his own personal fortune.

IC^2 founded organizations, such as the Austin Technology Incubator (ATI), The Capital Network, and the Austin Technology Council. These ongoing activities continue to contribute to the Austin region's technology based growth, and serve as "experiential learning laboratories" nationally and internationally. In addition, IC^2's other core programs include IC^2 Institute Fellows, Commercialization Training and Consulting, the Masters of Science in Science and Technology Commercialization (MSSTC) degree, and Visiting Scholars. These programs serve to expand the Institute's international network of talent and educational outreach. IC^2's "new experiments," such as the Clean Energy Incubator (CEI), Cross Border Institute for Regional Development (CBIRD) and the Digital Media Collaboratory (DMC), continue to apply the Institute's vision in new ways, improving quality of life and promoting civil society at home and abroad.[30] Also, the Austin

Software Council was established by the IC² Institute to support the growing Austin software industry. It is now an independent entity.

The IC² Institute is a pioneer in technology-enabled education for working professionals. The Institute was the first to establish a MS Degree in Science and Technology Commercialization (MSSTC), a graduate degree from UT designed for working professionals. IC²'s MSSTC Program was the first UT degree to be offered simultaneously in Austin and outside of Texas (in the Washington D.C. area, and a certificate program in Lisbon, Portugal). The MSSTC Program is a UT's pioneer in the use of synchronous, technology-enabled learning and distance education. Learning modules from the MSSTC Program have been shared with other UT system programs, as well as delivered to students in Latin America via telecommunications technology. IC² has active strategic alliances with leading academic institutions in 25 nations via its fellows and partners. The Institute has sponsored more than 50 UT graduate students at these institutions since 1990.[31]

The IC² Institute pioneered the concept of the "Technopolis Framework." The framework emphasized interlocking relationships between academia, business and governments to foster technology-based economic development.[32] A paper co-authored by the Institute's fellows outlined the four fundamental steps required to develop a region as a technopolis:[33]

- Achievement of scientific pre-eminence in technology-based research
- Development of new technologies for emerging industries
- Attraction of major technology companies
- Creation of homegrown technology companies

Austin is a prime example of this technopolis framework in action.

2.2.4. The Austin technology incubator

By the late 1980s, Austin was mired in a severe recession, signified by its empty office space and high unemployment. Yet the capital city still had a university with 50,000 students and major research activities; the world-class research facilities of IBM, Motorola, MCC and SEMATECH, which provided career opportunities for a young and highly educated workforce; a high quality of life that attracted increasing numbers of people from outside the city; and an increasingly motivated city leadership who was diligently searching for tangible solutions to the region's economic woes.

Against this backdrop, the Austin Technology Incubator (ATI) was formed by a coalition of university, government and business leaders led by George Kozmetsky. The City, the County, the Chamber of Commerce and Kozmetsky funded a three-year experiment to create wealth, generate jobs, diversify Austin's struggling economy, fill office space and build an entrepreneurial infrastructure for the city. The experiment met success with the recruitment of three promising technology start-ups: one from California, one from UT and one from the Microelectronics and Computer Technology Corporation (MCC).[34]

The ATI was founded by the IC^2 Institute at UT in 1989. At the time Austin was emerging as a technology center, with the successful recruitment of major divisions of IBM, Motorola and AMD; after winning a major MCC national competition in 1983, Austin became home to major research divisions of 3M and SEMATECH.[35]

The ATI specialized in helping young companies that did not know how to translate a good idea into a successful product. ATI helped shorten product development cycles by broadening tenant entrepreneurs' know-how in market research, finance, sales and service and other elements critical to successful entrepreneurship. They provided tangible items like phones, Internet access, fax machines, video-conferencing, etc. However, according to ATI officials, what added the most value were intangibles: an entrepreneurial culture, community partnerships with city government, relationships with other companies and consulting services, and links with venture capital networks.[36]

Furthermore, ATI in conjunction with IC^2 created The Capital Network (TCN), a non-profit seed capital network. TCN was designed to stimulate seed capital and venture capital opportunities for Texas entrepreneurs. TCN is now an independent entity and is the largest and most successful seed capital network in the United States.[37] ATI and TCN were important catalysts for attracting talent, raising the venture capital that motivated the talent to come to Austin.

The efforts that each of these institutions put into making Austin a technopolis began to bear fruit, slowly at first and then dramatically during the 1980s. By the late 1990s, Austin was comparable in amenities to any of the country's major cities.

2.3. Human resources

Austin has a growing population of highly skilled workers. According to the U.S. Census in 2000, the Austin metropolitan and surrounding areas have a population of 1.25 million people in about 11,000 square kilometers.

Austin's population has been growing at staggering rates since the 1980s – the population growth in Austin and its surrounding counties being in double digits. The population of Williamson County nearly doubled between 1980 and 1990, while the surrounding counties took two decades to double in size. During this time the population growth for the state of Texas was around 20 percent, while the Austin area was experiencing growth of more than 40 percent during this ten years. In comparison, growth for the U.Swas just above 10 percent during the same decade. By 2004 the Austin area population had grown to more than 1.4 million people. In addition, Austin attracts a younger population than the national average, primarily in the 25–44 age group. UT, with its 50,000 students, helps to attract this younger population.

UT is both a source for highly skilled recent graduates and a magnet for attracting highly skilled mid-career professionals. Since 1989, UT has consistently ranked in the top 10 engineering graduate schools in the United States, according to *U.S. News and World Report*.

In 1995, the College of Engineering created a program called the Executive Software Engineering Master's Degree (also known as the Option III Master of Science in Engineering Degree), with an emphasis in software engineering, intended for working professionals in the computer industry. Also significant for the ongoing development of software professionals in the Austin area was the Software Quality Institute (SQI), which started in 1993 as part of the College of Engineering. SQI courses are taught by industry practitioners and focus on management training for professional certifications.[38]

Austin's economy added 163,900 jobs between 1990 and 1996, a 5.3 percent compound annual growth rate, while the population grew by 200,000. Austin created 21,600 new jobs in 1996 alone and added another 44,000 in 1998, a growth rate of four percent a year. In 1996, Austin added one out of every 10 new jobs in Texas, while creating one out of every five manufacturing jobs.[39]

In 2004, Austin boasted more than 30,000 firms employing more than 710,000 people in business and manufacturing. The average annual income for a business professional was nearly $40,000, while the average annual income in manufacturing was about $64,000.[40] The unemployment rates of Austin and Travis County was consistently lower than those for the State of Texas and for the U.S. as a whole. During the 1990s, Austin's unemployment rates were significantly lower than the state and national rates.[41]

Austin has earned its reputation as a city of creativity, innovation, and invention. Patents are one of the best measurements of innovation, and Austin has become competitive in the number of patents issued yearly.[42] In 2004 the number of patents awarded to inventors in Austin was 2,069, a dramatic increase from the 74 patents awarded in 1975. It is significant that the number of patents from Austin grew more than 27 times during a period in which the total number of patents filed in the United States did not quite double.[43]

UT still lags behind other top research universities in patent filing and licensing. In 1995, UT earned $1.2 million in patent licensing revenues compared to $5 million for MIT and $39 million for Stanford. In terms of the number of patents issued in 1995, UT ranked fifth with 26, far behind both MIT and Stanford. In biotech industries, the UT system ranked tenth in the cumulative number of patents held by all institutions. UT officials expressed confidence that licensing revenues would spike upwards, since the university had substantially increased the size of its Technology Licensing & Intellectual Property Office in 1997. According to the Greater Austin Chamber of Commerce, in 2004 UT was third in the nation in the number of patents earned.[44]

During 1990 to 2004 the total number of patents awarded to Austin inventors was 20,066. According to a U.S. Patent and Trademark Office geographic analysis of patent activity in the 1990s, the number of the city's patented inventions was comparable to that of Dallas, San Diego, and Atlanta. Austin outpaces much larger and more economically diverse regions such as Phoenix, Seattle, Denver and Raleigh-Durham.[45] However, Austin trailed Boston's Route 128 and Silicon Valley in major research centers and university research and development expenditures.[46]

Another factor helping Austin has been the U.S.'s relatively open immigration policies for students and high skilled workers, which has allowed U.S. universities and companies to attract the best and brightest students and workers from around the globe. According to the annual report on international students published by the Institute of International Education, with support from the State Department's Bureau of Educational and Cultural Affairs, there were 572,509 international students enrolled in U.S. higher education institutions in 2003–2004.[47] The report ranks UT Austin fifth in the nation, with 4,827 international students. Furthermore, the U.S. Department of Commerce data ranks higher education as one of the top 10 largest service sector exports for the United States.

Studies in the last 50 years have shown that immigrants help to revitalize the U.S. and create economic stability within communities. A report issued by the Presidential Commission on Immigration and Naturalization in 1953 noted that "[t]he richest regions are those with the highest proportion of immigrants. Their industry, their skills, and their enterprises were major factors in the economic development that made these regions prosperous."[48] While data is not available on the effects of highly skilled immigrants in building the technology cluster in Austin, it is clear that Austin attracts the best and brightest workers from around the world.

2.4. Living conditions

In 1838, Mirabeau B. Lamar, sent to scout for a site for the capital of the new Republic of Texas, described the site he found on the Colorado River as "the seat of a future empire."[49] Austin became the capital of the Republic of Texas and later the capital of the state of Texas. It is now not only the seat of political and governmental power, but also the center of knowledge, technology and innovation. Young professionals find many employment opportunities at technology industry giants such as Dell, AMD, 3M, IBM, Samsung, Austin Semiconductor, and National Instruments. And on top of this, Austin is known for being the most creative, "rockin'," young, educated, physically fit, and creatively distinctive (many say "weird") city in Texas.

In 1996, Austin ranked number one among U.S. cities in job creation, and was considered to have one of the most highly educated populations in the country. Austin's unemployment rates have consistently remained lower than both the state and national averages. In 2004 the percentage of households in Austin with annual incomes above $75,000 was higher than the rest of the country. Additionally, the percentage of household incomes in the lower brackets was lower than the national average.

In 1996, Austin was ranked in the top 10 list of *Fortune* magazine's best cities, and was ranked eleventh in the "Lowest Cost of Living in a Major City" category.[50] In 2004 the cost of living in Austin remained one of the lowest among big cities. The standard of living, as measured by Austin's household income adjusted for the cost of living, led all comparable regions in the United States.[51] It is significant to note that Texas is different from most states in that it does not have a state income tax, but relies on sales taxes, property taxes, and other taxes.

The combination of higher household incomes and lower cost of living, home prices, and taxes generally results in young Austin professionals having a significant amount of "disposable income." This disposable income can be (and often is) spent going out to clubs, bars and concerts; purchasing cars and boats; remodelling, redecorating and home improvement; eating out at restaurants; and buying household items such as big screen televisions, surround sound stereo systems, computers, and other technological gadgets. Subsequently, Austin has more restaurants and clubs per capita than any other U.S. city.[52]

Austin has been named the "Best Place for Business and Careers," the "Best City for Relocating Families," and one of the "Best Cities for Singles." Also known as "the Live Music Capital of the World," Austin boasts more than 120 live music venues.[53] Each year Austin hosts the South by Southwest Music, Film and Media Conference & Festival, the Austin City Limits Music Festival, and the Austin Film Festival.

In addition to the live music, Austin is also know for its multitude of outdoors activities and for having a young, active, physically fit population. The city's location on the Colorado River allows residents to enjoy outdoor activities in and around the area. Austin's surroundings boast lakes for boating, sailing, water skiing, scuba diving, canoeing, kayaking, and trails for walking and running. Austin also has nearly 50 golf courses, trails for hiking and mountain biking, spring-fed swimming pools, bicycle racing, 5- and 10-K events, rock climbing, botanical gardens, a wild flower center and local, state and national parks. In June 2005, Austin was listed as one of the 19 most nature-friendly areas in the country by the book *Nature-Friendly Communities*, published by Island Press.[54]

In a series of articles published in the spring of 2002, journalists at the *Austin American-Statesman* enlisted several social scientists to help them identify the factors that fueled Austin's development. They determined that Austin was one of a new group of cities whose growth was being driven not by cheap labor, traditional manufacturing, or proximity to trade routes, but by a group of people called the "creative class." These were not only artists, musicians and writers, but software engineers, architects and scientists. In short, these were the people who developed new ideas.[55]

Austinites pride themselves on their creative and sometimes unusual culture. The popular slogan "Keep Austin Weird!" embodies the general sense of creativity, individuality and uniqueness of Austin's residents and neighborhoods. To this end, Austin has festivals throughout the year celebrating arts, music, holidays, pet animal races, beer, and a fictitious cartoon character's birthday, to name just a few.

Austin's climate is made for nearly year-round outdoor enjoyment and recreation. The temperatures during the winter are usually quite pleasant, ranging from freezing to a balmy 70^0 F (21^0 C). However, during the summer the temperatures are typically in the high 90^0s F (35^0–36^0 C), with readings above 100^0 F (37^0 C) occurring frequently during June, July and August. The Austin Hill Country tends to be more humid and have more lush plant life than the flat prairies of west Texas. Spring and summer can bring intense, heavy rains and occasional tornadoes.[56] Austin typically has 115 clear days, 114 partly cloudy days, and only 136 cloudy days per year.[57]

The quality of life and the need to be attractive to young talent was not as prominent an issue during the 1980s. Austin's signature approach evolved in the latter part of the 90s and then became a prominent feature of Austin's recruitment and development strategy in parts due to Kirk Watson's collaboration with community and business leaders.

3. Anchor firms

The Austin technology cluster is unique in the fact that it has had multiple anchor firms over the last 50 years. As the dynamic story of the development of the Austin cluster has unfolded, the anchor firms have also changed and adapted.

The first technology anchor firm in Austin was Tracor, Inc., a defense technology company that was started in Austin in the 1950s. Through the company's research and development work it spun off more than 25 other Austin-based technology companies. During the 1960s and 1970s, IBM and Motorola were arguably the anchor firms in Austin, attracting numerous other companies and suppliers for their electronics manufacturing plants. In the early 1980s, MCC's choice of Austin as its headquarters brought worldwide recognition of Austin as a technology cluster. Consequently, more technology companies began moving to Austin. Similarly, when the SEMATECH consortium of semiconductor companies moved to Austin the city attracted more semiconductor fabrication and research facilities. At present, Dell is Austin's largest computer technology company, attracting other technology firms and suppliers of computer parts and accessories, and fostering entrepreneurial development of new technology start-up companies. This paper specifically examines the three most influential anchors in Austin: Tracor, MCC and Dell.

3.1. The Tracor case

The critical role of the research university in the development of a technopolis can be effectively demonstrated through a case study of Tracor, Inc. This is a homegrown company that was the first *Fortune 500* company headquartered in Austin. Tracor is an example of what has been called a high-innovation company, an incubator organization and what this paper refers to as an anchor firm.[58]

Frank McBee, the lead founder of Tracor, earned both his bachelor's degree in 1947 and his master's degree in 1950 in mechanical engineering at UT, after serving as an Army Air Corps Engineer from 1943–1946. In the late 1940s, McBee became an instructor and then an assistant professor in the UT Department of Mechanical Engineering. In 1950, he became the supervisor of the mechanical engineering department of UT's Defense Research Laboratory, now called the Applied Research Laboratory, at UT's Balcones Research Park.[59]

In 1955, with funding of $10,000, McBee joined forces with three UT physicists to form Associated Consultants and Engineers, Inc., an engineering consulting firm. Drawing on UT training and work experience, the four scientists focused their efforts on acoustics research. They were awarded a $5,000 contract for an industrial noise reduction project. The company's name was changed to Texas Research Associates (TRA) in 1957. During the late 1950s, the four scientists taught and did research at UT while working on developing TRA. In 1962, the firm merged with a company called Textran and adopted its present name of Tracor, Inc. At this time, McBee left UT to devote his time to building the company.[60]

The College of Engineering and the Defense Research Laboratory at UT nurtured the educated talent that formed the entrepreneurial venture of Associated Consultants and Engineers in 1955, which led to the establishment of Tracor in 1962. However, even more impressive is the constant stream of entrepreneurial talent that came from Tracor itself. More than 25 companies have spun off Tracor's research and development since 1962 and have chosen to locate in Austin, including Pison Associates, Inc.; Unitech; Texas Computer Corp.; Radian; Tracoustics; Espey-Huston; Tanga, Inc.; AMI; Spenco Systems; Guerrero's Photographic Group; Texas Telesystems; Texas Research Institute; SGE, Inc.; DAC International; Continuum Co.; Weed Instruments; and Key Concepts.[61] In 1998, General Electric Company acquired Tracor for about $1.4 billion.

Tracor and its spin-offs have had a dramatic impact on job creation in Austin. A total of 5,467 persons were employed in these companies in 1985. These companies are also capable of creating spin-offs of their own. Radian Corporation, for example, has spun off four companies.[62]

3.2. The microelectronics and computer technology corporation (MCC)

MCC formed in 1982 as a consortium of 10 high-technology companies in the computer industry, with the intent of preserving America's dominance of the computer industry, mostly in response to developments in the electronics and technology industries in the Asia-Pacific region.[63] The consortium's first CEO, Admiral Bobby Inman, led a nationwide search among 57 cities for one in which to locate MCC. Several of MCC's member companies, such as AMD, Texas Instruments, Motorola, Lockheed, and National Semiconductor already had branch plants in Austin, which was one factor in the consortium's decision.[64]

The location of MCC in Austin helped to bring other high-tech companies to the city. MCC chose Austin as its headquarters in 1983 after a major public site selection process among some of the most visible high-tech centers in the United States. As a result, Austin made headlines in the New York Times, the Wall Street Journal, and the world press as the next great "Silicon Valley." Austin was nicknamed "Silicon Prairie," "Silicon Gulch," and "Silicon Hills," and experienced an unprecedented wave of enthusiasm because of the perception that it had suddenly become a major technology center.[65] When SEMATECH, a consortium focused on increasing U.S. competitiveness in semiconductor tooling equipment, was looking for a home five years later, it was easy to make the case for Austin, partly because of the prominence received by attracting MCC. This in turn drew Applied Materials, Silicon Valley's leading manufacturer of semiconductor equipment, to locate a facility in Austin, which had the effect of luring numerous smaller suppliers to Austin as well.

3.3. Dell, Inc.

Dell, Inc. is Austin's most visible success story to date. It started in 1984 when Michael Dell began making PCs to order out of his garage while a student at UT. Dell became Austin's first homegrown multibillion-dollar company and Austin's largest private sector employer.[66] Dell currently employs more than 5,000 people in Austin and nearly 30,000 people worldwide. Since its Initial Public Offering (IPO) in 1988, Dell has seen its revenues grow dramatically, reaching an estimated $50 billion in 2005.

Dell's worldwide market share has increased six-fold in the last ten years. Despite a slump in consumer demand, the company's sales outside the U.S. rose 20 percent during the third quarter of 2005 compared to the same period of the previous year. This accounted for 40 percent of Dell's worldwide revenue in the quarter. Asia Pacific and Japan saw a 20 percent year-over-year growth, while China, a key market for Dell, saw unit growth of 46 percent.

4. Supporting firms, institutions, and individuals

4.1. Cooperation and competition between related firms

The technopolis wheel framework makes it clear that the role of key individuals or influencers in facilitating interaction among the seven major segments of the wheel is critical to developing a technopolis cluster. The interaction among the segments fosters new institutional linkages among business, government and academia, which may promote economic development and technology diversification.

As a result, a fascinating paradox has emerged – the paradox of competition vs. cooperation. On the one hand, a great deal of competition takes place between a state's universities, companies, and public and private sector entities. On the other hand, cooperation is essential for a technopolis to develop and survive over time. The UT study emphasizes the importance of networking across the seven segments of the technopolis wheel. It argues that in order to initiate a change it is necessary to have the ability to link public and private sector entities, some of which have been traditional adversaries.[67] This study specifically highlights the role of UT and the role of cooperation among local government and support groups as critical factors in Austin's development.

4.2. Role of the research institution

UT, like any public university, competes for state funding with all of the other state agencies, programs, services and institutions. This is the classic economics problem of scarce resources. In 1984, shortly after the MCC decided to locate in Austin, oil prices were still at about $30 a barrel, and state revenues increased by $5.4 billion or 17 percent over the previous year. However, the State of Texas decreased appropriations for higher education by three percent in 1984.

UT also must compete with other universities and research institutions for distinguished faculty and research fellows. Despite UT's phenomenal growth in endowed chairs, professorships, lectureships and fellowships, despite the location of MCC in Austin, and despite the national and international press citing UT as a new center of excellence in education, the lack of sustained state support for higher education sent a mixed message to the best scholars and researchers whom the university was trying to attract.

In summary, as state allocations for higher education increased through the late 1970s and the early 1980s, the perception of the development of Austin as a technopolis outside the state increased proportionally as well. On the other hand, as the State of Texas began to reduce higher education funding in the 1983 session, the perception of Austin as a developing technopolis declined and the perception of retrenchment in the university began to emerge.[68]

In part, the reduction in state funding led UT President Cunningham to begin to look more for outside sources of funding. The Bayh-Dole Act made it possible to create profitable university-industry research collaborations and to make profits from patents resulting from federally funded research.[69]

4.3. The role of government

Federal, state and local government play vital roles in the development of a technopolis. However, each level of government affects economic development differently. The federal government has had an impact on the development of the Austin cluster in five key ways.

First, the development and operation of Bergstrom Air Force Base, established in 1942, provided fundamental economic stimulation to the Austin region through the employment of 1,000 civilians and 6,000 military personnel with an annual payroll of about $167 million, according to U.S. government documents. Bergstrom Air Force Base closed and subsequently became Austin Bergstrom International Airport, which has been a major infrastructure development that has helped Austin's growth as a technopolis.

Second, federal funding for research and development activities at Balcones Research Park at UT constituted direct government stimulation of the emerging Austin technopolis. The Balcones Research Park was created in the early 1940s when the federal government provided the land for UT- funded research in strategic resources to support the war effort in World War 2.[70]

The third significant government factor that fostered the growth of technology research at UT was the passage of the Bayh-Dole Act. Under this law, universities and businesses could obtain exclusive licensing of new inventions after they committed to diligently developing and transferring the invention to the market for the public good. One effect of the act was to foster university-industry research collaborations.[71]

Fourth, changes in federal tax laws in 1979, 1981 and 1986 pertaining to capital gains encouraged investments in venture capital pools. This coincided with a dramatic growth in venture capital investments, both in Austin and around the country. Austin is currently ranked fourth out of U.S. cities for venture capital investment, with about $540 million dollars.

Fifth, the federal government's immigration policy changes have affected economic development by offering more student and work visas and by raising or eliminating quotas. One way in which the U.S. government encourages high-skilled foreign nationals to work in the U.S. is through temporary work visas; another way is through student visas. The State Department states that there were 572,509 international students enrolled in U.S. higher education institutions during 2003–2004.[72] UT ranks fifth in the nation in numbers of international students, there being almost 5,000. Being able to attract the best and brightest students and professionals from around the world is extremely important for research and development, patent writing, and entrepreneurial activities.

The state government's primary role has been to provide funding for higher education and tax incentives for businesses. Governor White proved the importance of visionary state leadership when he made winning the competition for MCC a top priority. The local government's primary role in Austin has focused on competitive rates for utilities and infrastructure requirements, such as roads, and on boosting quality of life. In Austin, a high quality of life has remained relatively affordable in comparison to other technology centers.

Texas has re-awakened to the need to play the incentives game since 2004. At the same time, New York was offering $1 billion in incentives to build the nanotechnology and semiconductor industry in New York. When New York tried to recruit SEMATECH away from Austin, Texas leaders collectively tried to keep SEMATECH in Austin. They were successful, and that effort lead to the creation of the Texas Enterprise Fund, which recruits companies to Texas, and the Emerging Technology Fund which funds university-related commercialization.

4.4. The role of support groups

Perceptions of the desirability of growth often vary within any region undergoing rapid economic growth, and there is always the possibility that such growth will diminish the very qualities that caused the area to be so attractive to companies in the first place. The rapid growth associated with the development of the Austin technology cluster is no exception. Tensions between a sustained quality of life and sustained economic development have been visible throughout the development of Austin through election outcomes and community groups representing environmental concerns, labor issues, minority viewpoints, and other community interests.[73]

Support groups can provide an important mechanism for the development of a technopolis or cluster. These groups may take a variety of organizational forms representing research institutions, entrepreneurial support within institutions or companies, incubators, venture capital investment firms, and community interests. Business-based groups, such as accounting firms, law firms and major banks, provide necessary services and incentives for high-technology companies in an emerging technology cluster. These components provide a source of expertise and a reference resource for those founding and/or running technology-based enterprises.[74]

4.5. Institutional entrepreneurial support

In 1998, the companies that were listed as major benefactors of the College of Engineering included some of the big names in the technology industry, such as Intel, Motorola, Texas Instruments and AMD. However, virtually none of the benefactors were companies founded by recent UT Alumni who had become successful entrepreneurs. In comparison, Fred Terman at Stanford had encouraged his students in their early product development efforts in the 1930s. Texas was hampered by the fact that it had strong policies disallowing the use of state property for commercial use. Some companies such as Samsung had addressed this issue through formal research contracts with the university that allowed them to fund research that had both commercial and academic goals. However, this option was not yet available to cash-starved start-ups.[75]

In 1998 UT professor Ben Streetman, who had helped to lure MCC, was coordinating efforts with the Austin Software Council to introduce UT students to several local start-up companies, rather than relying on the traditional recruiting efforts of large corporations.[76]

The UT Graduate School of Business (GSB) had, over the years, made strenuous efforts to promote entrepreneurship in Texas. This was largely the legacy George Kozmetsky and the relationships he continued to foster after forming IC^2, the ATI, and The Capital Network venture fund. The GSB offered four courses in the MBA program that focused on entrepreneurship and sponsored a competition known as the "Moot Corp" to encourage entrepreneurship among students.[77] The national and international Moot Corp competition, now administered by the UT GSB, was created and is supported by IC^2 as a catalyst for entrepreneurship among Texas graduate students.[78] The goal of the competition was and is to boost the spirit of entrepreneurship among students at UT. *Business Week* dubbed it the "Super bowl of Business Plan Competitions," as it attracted teams from business schools around the world.[79] Winners of the competition were given free housing at ATI for a two-year period. In 1997, Dean of the College of Business Administration Robert May, announced a plan to establish a "Center for Entrepreneurship, Growth and Renewal" at UT. In 1998, *Success* magazine ranked UT's GSB in its top 10 list of schools for entrepreneurship, ahead of Harvard, Babson College, Stanford, MIT and UC Berkeley.[80]

Major technology corporations have found various ways to encourage entrepreneurial activities among their employees, through corporate "intraventures" or "new initiative groups," venture capital investment, spin-offs and internal and external incubators.

4.6. Incubators

In the U.S., about 75 percent of all incubators are non-profit entities that are supported by local governments, academic institutions of higher learning and/or local businesses. In 1998, a report issued by the NBIA placed that number at more than 90 percent. During the ".com" mania of the late 1990s there was a pronounced shift towards for-profit incubators; however, the decline of stock markets that began in March 2000 signaled the demise of many of these Internet-focused incubators.

During the past few years, large corporations have developed several models of for-profit incubation, sometimes referred to as corporate "intraventures" or "new initiative groups." Some corporations, such as Intel, Dell and HP, have set up corporate venture funds that invest in start-ups outside the corporate walls. Others, like Samsung, spin off technology teams to be incubated offsite. Still others like UPS, Panasonic, Adobe, Ford, Lucent and Trilogy have developed more traditional incubators.[81] Consultancy companies and venture capitalists, such as Austin Ventures, Benchmark Capital and Kleiner Perkins Caufield and Byers, have also entered the incubator market.[82]

The Austin Technology Incubator (ATI) was founded by the IC^2 Institute at UT in 1989. ATI was formed by a coalition of university, government and business leaders led by George Kozmetsky. The City, the County, the Chamber of Commerce and Kozmetsky funded a three-year experiment to create wealth, generate jobs, diversify Austin's struggling economy, fill office space and build an entrepreneurial infrastructure for the City. The experiment became a success with the recruitment of three promising technology start-ups: one from California, one from UT and one from MCC.[83]

Over the years, ATI has fostered the growth of more than 140 high-tech companies, graduated more than 75 companies, created more than 3,000 jobs, launched five IPOs and 13 acquisitions and generated more than $1.4 billion in revenue.[84]

4.7. Venture capital

The growth of venture capital provides a good example of the importance of business-based groups to the development of a technopolis. In 1980, Austin had virtually no venture capital money. However, by 1986, the city had about $80 million managed by five firms. The growth was due primarily to two factors: changes in federal tax laws, and changes in perceptions of Austin. The changes in federal tax laws in 1979, 1981 and 1986 pertaining to capital gains encouraged investments in venture capital pools, while the perception of Austin as an emerging technology center encouraged the development of homegrown pools. Venture capitalists in Austin, while wanting a local window on technology and company development, did not see enough good deals, i.e. fast-growth company potentials, in the region. The sources of the venture capital were a few individuals knowledgeable about the venture capital process, as well as the major commercial banks in the area. Although funds in these pools increased, most venture capital investments in the country continued to be made outside the state of Texas.[85]

The Austin Technology Incubator (ATI) helped create The Capital Network, a non-profit seed capital (angel) network and the largest network of its kind in the United States. Venture funding in Austin in 2004 included about $539 million that went to local companies. This is more than 12 times higher than the 1995 amount of $42.4 million. However, the story is not one of uninterrupted success; the amount of venture funding in Austin spiked in 2000, but by 2004 it was only about one-quarter of its bubble-era peak. It is important to note that in 1996, Texas ranked fourth in the nation in terms of total venture capital funding activity, behind California, Massachusetts and Florida.[86]

Although the establishment of The Capital Network is a step in the right direction, the Capital Area Network's sustainability is perhaps more important for the healthy growth of new technologies in future. It would be useful to conduct further studies of how the network emerged, the role it served, the mistakes it made, and, ultimately, how it outlived its usefulness.

During 1996 and 2004, investments originating in Austin were only a very small faction of the total investment in the region. The total venture capital invested in Austin Metropolitan area companies between 1995 and 2004 was $6.4 billion for a total of 654 companies. In comparison, during the same years, the total venture capital investment for all U.S. companies was $315.4 billion for 30,604 companies nationwide.

Austin's oldest VC firm was Austin Ventures, which had $448 million under management in 1998. Although Austin Ventures had participated in several non-Austin start-ups such as SynOptics, a Silicon Valley company, its focus was on companies based in the southwestern United States and Austin in particular. Some of Austin Ventures' Austin-based firms that had done well were Tivoli Systems, which went public in 1994 and was bought by IBM a year later for $743 million, and Crystal Semiconductor, founded in 1984 and sold to Cirrus Logic in 1990.[87]

MCC's selection of Austin had focused a spotlight on Austin as a place where large technology companies could set up shop and be assured of conditions where they could flourish. Principals at Austin Ventures believed that a spate of successful IPOs in Austin would create a similar splash, showcasing Austin as a place where entrepreneurs could succeed. Dell's success in creating "Dellionaires," or people who made millions of dollars from Dell stocks and who then created their own high-tech companies, was an example of what Austin Ventures was trying to accomplish through the companies they funded and eventually hoped to take public. Conspicuous wealth creation, they hoped, would act as a magnet for scarce executive talent to migrate to Austin in search of companies that would give them an opportunity to get rich through a successful IPO in exchange for a few years of hard work. Austin Venture's ambition was to convert Austin into a "small company town" rather than one dominated by a few behemoths, and into a destination for the entrepreneurial-minded.[88]

Some people believe that Austin had beaten Silicon Valley as a place for entrepreneurship to thrive. Other Austinites were more cautiously optimistic. Austin business leader Pike Powers believed that Austin would eventually do one better than Silicon Valley, because it had Silicon Valley to learn from. He believed strongly that Austin offered a much "richer brew," in terms oflifestyle, diversity of its environment, and in the type of people who chose to make the city their home, than other cities and regions such as Silicon Valley.[89] Yet others warned of the lack of role model entrepreneurs whose experiences could serve as lessons to other budding entrepreneurs.

4.8. The role of the private sector

4.8.1. Private sector partnerships

Private sector partnerships can be useful, particularly in areas like technology where research and development are taking off like wildfire. One example of such a partnership occurred in the early 1990s, when IBM and Motorola decided to locate their joint effort, known as "Somerset," to design the PowerPC microprocessor in Austin. IBM and Motorola intended the PowerPC to challenge Silicon Valley-based Intel Corporation's dominance of the PC microprocessor industry. AMD's K-5 microprocessor, which was conceived and designed in Austin, was intended to challenge Intel's virtual monopoly of PC microprocessors Thus, at one point, Austin was home to the two major efforts in the semiconductor industry to "dethrone" Intel from its number one position in the PC microprocessor market.[90] Intel set up its own design center in Austin in 1998.

4.8.2. Consortiums

MCC and SEMATECH are examples of consortiums consisting of traditionally competitive companies that decided that a certain level of cooperation was in the best interests of all companies. The MCC consortiums grew out of the need of U.S.-based technology and semiconductor companies to ensure their competitiveness with foreign-based companies. MCC began as a consortium of 10 major computer and semiconductor manufacturers that was formed in 1982, in response to Japan's Fifth Generation initiative, to preserve America's dominance in the computer industry.[91] Although MCC was a key factor for Austin's technology industry recognition nationally and internationally, it ceased to operate about 10 years ago when it successfully met its goal by re-establishing US dominance in the computer industry.

SEMATECH was a bold experiment in industry-government cooperation that was conceived to strengthen the U.S. semiconductor industry's competitiveness in semiconductor tooling and equipment. The consortium, called SEMATECH (SEmiconductor MAnufacturing TECHnology), was formed in 1987, when 14 U.S.-based semiconductor manufacturers and the U.S. government came together to solve common manufacturing problems by leveraging resources and sharing risks. The goal of the organization is to accelerate the commercialization of technology innovations into manufacturing solutions. It accomplishes this by addressing critical challenges in advanced technology and manufacturing effectiveness and by finding ways to speed development, reduce costs, share risks and increase productivity.[92]

4.8.3. Spin-offs

Many Austin technology companies spin off new technology startups. Two examples of such companies are the Trilogy Development Group, which provides software to automate sales and other business operations, and

Tivoli Systems Inc., which develops software to manage computer systems. Both companies have spawned multiple technology start-ups. It is interesting to note that Tivoli was a software company founded by former IBM employees, and which was eventually bought by IBM. UT has also played an important role in the development of the Austin technology cluster by spinning off companies. Out of 103 small and medium-size technology companies in Austin in 1986, 53 percent indicated that their origin had a direct or indirect tie to UT. These companies' founders were UT students, graduates, faculty members, and other UT employees.

4.9. The role of visionaries/influencers

Although each of the institutional segments in the technopolis wheel is important to high-technology company development, the ability to link or network the segments is the most critical factor. Indeed, unless the segments are linked in a synergistic way, then the development of the technopolis slows or stops. In Austin, influencers have been responsible for linking these segments – key individuals who make things happen and who are able to network with other influencers in each of the other segments as well as within each segment. For example, in the Austin cluster the key influencers include George Kozmetsky, former Dean of UT's College of Business Administration and founder of IC^2 and ATI; Pike Powers, who managed the city of Austin and Texas' campaign to attract MCC; Lee Cooke, former President of the Chamber of Commerce and Mayor of Austin; Mark White, Texas Governor; Herbert Woodson, Chairman of the EE Department at UT; and Ben Streetman, UT professor and one of the leading academic figures in the semiconductor industry, who helped to attract MCC to Austin. Influencers provide leadership in their specific segment because of their recognized success in that segment. They maintain extensive personal and professional links to all or almost all the other segments. They are highly educated. They move in and out of the other segments with ease, and they are perceived to have credibility by others in the other segments. Influencers who represent business, academia, government, and community interests facilitate cross-segment linkages.[93]

4.10. The role of the lower Colorado river authority

The LCRA first made it possible and safe to live in Austin by building the dams and creating the reservoirs that later allowed for tremendous population and industry growth. These dams provide services such as electricity, water supply and wastewater treatment. The LCRA ensured that there were water, electricity, and wastewater services to meet the needs of the growing population, while maintaining the environmental quality that first attracted many people to Austin. It also managed the reservoirs so as to supply water to the water-intensive technology fabrication and manufacturing plants. "Without the LCRA, Austin's growth, high tech or otherwise, would not have happened," said Cullick. "The Colorado River, without the dams and reservoirs, is an insufficient water supply for Austin's population and the repetitive flood damage would be too devastating to allow for continued growth."[94]

5. From agglomeration to cluster

Until the early 1980s, most efforts to promote industries in Austin were driven by the Chamber of Commerce. However, Austin largely remained a secondary platform for technology, rather than the place where ideas were generated and taken to fruition. The effort to change the focus of Austin into a high-tech center with a focus on homegrown innovation rather than secondary manufacturing began in the early 1980s. It coincided roughly with the time when the Microelectronics and Computer Technology Corporation (MCC) decided to locate its headquarters in Austin.[95] Momentum for the development of Austin as a technopolis reached a crescendo in 1983, when MCC chose Austin as its headquarters.

5.1. Bringing MCC to Austin

In 1982, Governor of Texas Mark White believed that Texas could not afford to continue its dependence on oil revenues and agriculture for its economic growth. He wanted to find a way to build human capital as a resource base for Texas' economic development.[96]

In early 1983 the Mayor of San Antonio, Henry Cisneros, first drew the governor's attention to the possibility of luring MCC to Texas. The governor personally led the Texas delegation to Chicago in March 1983 where three Texas cities, San Antonio, Dallas and Austin, made their pitch to the MCC site-selection committee.[97] Austin was the only city in Texas to make it on the final list. Governor White told Pike Powers, executive assistant to the Governor and former chairman of the board for the Austin Chamber of Commerce, that it was Powers' job to ensure that MCC came to Austin, and entrusted Powers with the job of working with UT and city and business officials to accomplish that goal.

Powers believed that he and Austin were lucky that most people saw MCC as crucial, not just to Texas, but also to American competitiveness in a critical industry. As a result, when MCC made Austin its home, it focused national attention on Austin. This was precisely the kind of major hit that Governor White had hoped to achieve.

In early 1988, after a national competition, the main players in the U.S. semiconductor industry chose to locate its own consortium, SEMATECH, in Austin.[98] It is important to note that Austin and Texas were outbid for SEMATECH by several other contending cities and states in terms of financial incentives. However, SEMATECH officials cited as Austin's synergy among business, academic, government and community entities as the main reason for choosing Austin over the other cities.[99]

Texas devoted significant resources toward recruiting MCC, as a branding effort and as a means of recruiting talent to Austin, and this project was a success while MCC existed. However, it should be noted that that MCC ultimately failed about 10 years ago. The failure of MCC points to the difficulty of making partnerships between commercial enterprises work. SEMATECH, is a more successful Austin consortium, but it too is facing major changes recently. David Gibson and Everett Rogers[100] provided insights to why these partnerships are so challenging. Although MCC did not survive in the long run, it left a long-term impact on Austin's high-tech industry by bringing a large amount of software and administrative talent to the regions. Even after MCC's failure, this talent remained in the region and continued to contribute to the technology industry of the Austin area.

5.2. Interaction of the technopolis wheel

The IC^2 study that identified the technopolis wheel used data from surveys, interviews and archival sources to assess the role and impact of each segment of the emerging Austin technopolis.[101] Analysis of the data demonstrates the role of the research university in forming spin-off companies, the direct and indirect impacts of federal, state and local government, and the evolution of high technology over time. The relocation of major companies to an area and the establishment of indigenous high-technology companies were found to be especially important. The study points to new concepts of institutional relationships among the segments of the technopolis wheel. It emphasizes the role of influencers, who provide leadership in each segment while networking the different segments to form new institutional alliances. Three important themes emerge in technopolis development, not only in the United States, but also in Europe and Asia: the need for a coordinated approach to high-technology company development, the presence of a high-quality research university, and the importance of a network of influencers or "executive champions."[102]

5.3. Building a new wireless cluster

The IC^2 Institute published a report in January 2004, titled Austin's Wireless Future, which found that there are over 91 wireless companies in Austin. The findings of the report surprised local business and community leaders, who had not previously recognized the growth in wireless companies and technologies in the Austin area. As a result, Austin leaders are looking into the future growth potential of the wireless industry, and are trying to make Austin into a wireless technology cluster. Austin has many of the elements required to build a wireless technology hub, including a major university with a significant wireless research program, local availability of raw materials and workforce, highly-developed Internet presence, and high degree of cultural creativity. Time will tell if Austin can successfully utilize the same techniques in creating a wireless cluster in the twenty-first century that worked to build its technology cluster in the twentieth.

6. Conclusions

6.1. Summary of key findings

There were many factors that influenced the development of the Austin technology cluster. Due to the high level of cooperation and coordinated efforts among different agencies and institutions, it is difficult to separate and prioritize the contributions of each. However, there are two essential factors that are worth highlighting. First, cooperation among local government and support groups was critical to Austin's development. Second, the university provided the core resource for attracting and developing "knowledge based industry" in Austin. Additionally, a number of key findings emerge regarding the development and maintenance of a technology cluster from the study of Austin, Texas. They are summarized as follows:

Austin is not the best at any one single factor in technology cluster development, but is good at many of them.

Employment in Austin has grown around technology-based companies.

Major technology firms have been attracted to and chose to locate in Austin for two primary reasons: 1) access to university resources, particularly the pool of skilled workers, and research and development opportunities, and 2) desire to operate in an affordable high-quality-of-life environment.

The research university has played a pivotal role in the development of the Austin technopolis by achieving scientific pre-eminence; creating, developing and maintaining new technologies for emerging industries; educating and training the workforce and professionals required for technology-based economic development; attracting large technology companies; promoting the development of home-grown technologies; and contributing to improved quality of life and culture.[103]

Visionary individuals played an important role in the development of the technology sector, particularly in the early years. These included Governor Mark White, George Kozmetsky, Lee Cooke, Pike Powers and the mayor of San Antonio. Note that the Kuchiki flowchart model does not take into account the role of visionaries in the formation of an industrial cluster.

Local government had a significant impact, both positively and negatively, on company formation and relocation, largely from what it has chosen to do or not to do in terms of quality of life and infrastructure.[104]

State government had a significant impact, both positively and negatively, on the development of the Austin technopolis through what it has chosen to do or not do for education, especially in the areas of making and keeping long-term commitments to fund research and development, faculty salaries, student support, and related educational development activities.[105]

The federal government played an indirect but supportive role in allocation of money for research and development programs and defense-related activities at the Balcones Research Park at UT; operation and closure of Austin Bergstrom Air Force Base; changes in federal legislation regarding private ownership of federally funded research; changes in tax laws regarding venture capital investments; and open immigration policy.

Continuity in local, state and federal government policies has an important impact on maintaining the growth momentum of a technopolis.

The establishment of activities and organizations to support entrepreneurial companies has been crucial, at UT, through IC^2, and through incubators such as ATI and venture capital investment firms including The Capital Network.

Large technology companies have played a catalytic role in the expansion of the Austin technopolis by maintaining relationships with major research universities, acting as a source of spin-off companies, and contributing to job creation. These companies have also served as an economic base that encourages related firms to relocate to the technopolis.

Small technology companies in Austin have helped in commercializing technologies, diversifying and broadening the area's economic base, contributing to job creation, spinning companies off the university and other research institutes, and providing opportunities for venture capital investment.[106]

Cooperation among traditionally competing companies, in the form of consortiums and joint projects and public-private partnerships, is also important to a technopolis. There must be cooperation and coordination among all segments of the technopolis in order to promote the development of an economic cluster.

State and local influencers have provided vision, communication, and trust crucial to developing a consensus for economic development and technology diversification, especially through their ability to network with other individuals and institutions in other segments of the technopolis wheel.[107]

Consensus or shared vision among and between segments of the wheel is essential for the sustained growth of the technopolis.

Austin's reputation for live music, entertainment and outdoors activities attracts and keeps young technology professionals in Austin. Austin is known for its live music, nightlife, and multitude of cultural activities and wide selection of restaurants.

Austin has a young, vibrant, educated, physically fit population. For the past several years Austin has ranked as one of the top cities for young professionals to live.

LCRA played a significant role in the development of Austin's infrastructure, helping to meet the needs of its growing population, including water supply, electricity and wastewater treatment capabilities.

Austin has developed other important forms of infrastructure, including building a new international airport and new roads.

Austin's warm climate and opportunities for outdoor recreation in the lakes and hill country promote a high quality of life.

The City of Austin and Travis County offer economic incentives that act as factors attracting technology industries to the region.

Austin still lags behind other technology cluster cities in venture capital investment.

Government bureaucracy was generally not a problem for Austin's economic growth. When obstacles did arise, there were politically powerful visionaries who were able to remove the hurdles before they could impede progress.

The open U.S. immigration policy allows highly skilled foreign nationals to immigrate for work. This kind of immigration is only easily possible in the U.S. and to a limited extent in Europe; contrast this with most Asian countries, where this is not the case. For example, someone from Japan could emmigrate to the U.S. and start his or her business and become successful, but the reverse is not easily possible due to closed immigration policies in Japan. Due to the open immigration policies of the U.S., Dell was able to succeed in Austin in ways that it could not in Japan or other Asian countries.

Austinites, and Americans in general, tend to value productivity over seniority. If one individual is more productive than someone who has more experience, the industry typically rewards the more productive person. This is generally not the case in Japan and other Asian countries, where seniority is valued more and it is often assumed that a senior person will be more productive.

The Austin technology cluster experienced a grassroots type of development. Most of the technology sector development in Austin happened because individual companies thought that Austin was a good place to move or start a company. This was not dictated by a governmental organization building an industrial zone. Again, this differs from the Kuchiki model.

Japan is to some extent indirectly responsible for Austin's growth. When Japan challenged the U.S. semiconductor industry, there was a sense of urgency among semiconductor companies to do something to keep their edge in the business. They accomplished this by developing new technologies at a faster pace than Japanese companies. Consortiums such as MCC and SEMATECH were formed as a part of this effort.

Austin celebrates its success stories, and through this celebration the word is spread to other businesses, which are encouraged to follow the trend and move to Austin. Some recent examples of companies that have "followed the hype" including Dell, Vignette, Tivoli and Trilogy.

Austin continues to show progress in the field of technological research and innovation, as evidenced by the number of engineering degrees granted and the number of patents received by individuals in Austin.

The Austin example further illustrates an interesting paradox: the very success of a developing technology cluster can lead to a decrease in the quality of life and community dissatisfaction.[108] For example, if the environmental effects of development and population growth were not addressed, the multitude of outdoor activities that attracted many people to Austin could be threatened. Often there is friction between advocates and adversaries of growth.

In conclusion, there is a need for continued cooperation among Austin's local government, university, businesses and community organizations. By 1998, Austin could look back on more than a quarter century of progress from a sleepy college town to a well-regarded center for technology research and manufacturing. Dell was the largest private employer in Austin, employing over 14,000 residents. Motorola was in second place, employing 12,000.[109, 110] Other Austin-based start-up companies such as Clear Commerce were on the cutting edge of Internet commerce. About 20 percent of all employment in Austin was attributable to the high-technology industry (up from just over 0 percent in 1970). Austin's unemployment rate had consistently beaten the national and regional averages for an extended period of time. Yet some Austinites were frustrated at Austin's lack of prominence in the world outside of the high-tech community. They repeatedly spoke of the need for Austin's institutions to cooperate and coordinate with each other if they wanted to duplicate California's Silicon Valley's success as a center for high-technology entrepreneurship.[111, 112]

Additionally there will need to be more cooperation in order for Austin to become a wireless technology cluster. A private economic consultant in 1998 said that the greatest danger Austin faced was "complacency." For Austin to achieve its long-term goals, "institutional commitments must be secured, renewed and strengthened among all of the players critical to Austin's success…All have a direct and critical goal in realizing Austin's future as a collaborative center for technology and knowledge advances for the nation."[113]

6.2. The Austin model

Kuchiki's flowchart provides a framework for understanding the development of an industrial cluster, based upon four key factors: (A) markets and industrial zones, (B) capacity building, (C) anchor firms and (D) related firms.[114] The first step in this flowchart model is (A), the government's decision to designate and build an industrial zone or export-processing zone, which in theory is based upon market factors for different types of industry. Building an industrial zone leads to (B), capacity building for infrastructure, institutions to support the new infrastructure and businesses, appropriately-skilled human resources to provide the labor necessary for manufacturing and/or research, and development and living conditions that attract human resources to the area. Once these items are in place, or at least in the works, they lead to (C), the attracting of an anchor firm to the industrial zone. The anchor firm (D) then attracts related firms, such as suppliers. Once enough related firms move to the area, the result is an industrial cluster. Figure 4.1 illustrates that this process flows thus: market forces → (A) → (B) → (C) →(D).

The Austin model is similar to this flow, but not exactly the same. In the Austin model, there is no equivalent to the industrial zone (A). While local and state governments can attempt to entice companies to move to an area that has been designated for commercial use or industrial manufacturing, the government doesn't typically develop or build the area. The development itself is left to the individual companies, if they decide to move there. Consequently, in the Austin flowchart, the anchor firms (C) begin a search for a location, such as in the case of MCC's nationwide search for a base city. Typically, cities and states compete with each other, providing plans and designs for (B), capacity building in the area, in order to meet the needs of the company or firm; in this case, the incentive was called (appropriately enough) the Texas Incentive for MCC.[115] When the company selects a location, this is when major capacity building activities typically begin, such as the building of infrastructure. Then (D), related firms move into the area to supply the anchor firm(s). Figure 4.2 illustrates that this process is: market forces → (C) → (B) → (D).

In the Austin example, the University of Texas at Austin falls under "capacity building," as it is an institution that has played a major role in the development of the technology cluster. It appears that the Austin case deviated from the typical Kuchiki model by giving more importance to infrastructure, institutions, human resources, and living conditions in terms of capacity building. In addition, it appears that the interaction of these four components, and the role played by key visionary individuals, was successful in facilitating the interaction necessary to reach the common goal of Austin's development. George Kozmetsky's Technopolis Wheel appears to describe the many factors, influences, and players involved when cities in the U.S., particularly Austin, compete for anchor firms; namely, the step between (C) → (B).

Key differences that emerge when comparing the Austin example with Kuchiki's flowchart model are as follows:

Visionaries' role is vital for Austin success. Influencers like George Kozmetsky, Governor Mark White, Pike Powers, Lee Cook, Kirk Watson and Henry Cisneros were able to draw global attention to Austin, transforming Austin from a sleepy college town to what it is today.

Networking is very important for Austin technology culture. This networking is very informal in Austin and very accessible, which may not

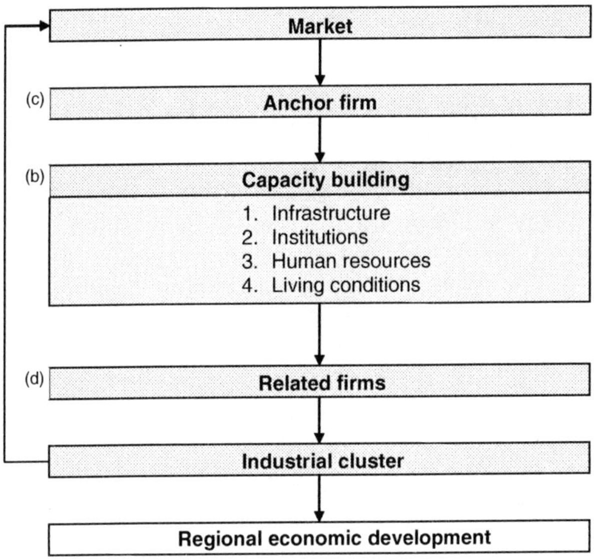

Figure 4.2 Flowchart for Austin technology cluster

Source: author.

be the case in formal and structured social settings outside of the Untied States.

No manner of seniority system was accepted and supported by Kozmetsky and others when encouraging young entrepreneurs. Such a meritocratic environment may be hard to achieve in formal and structured social settings like in Japan and China, but it seems to have great value in cluster creation.

In Austin, trying and failing is acceptable and does not bring dishonor to the individual or the family, which may not be case in several Asian counties.

The role of women in the success of Austin's technology industry is significant. This may not be the case in other societies in which traditional gender roles are entrenched.

The role of immigrants from outside the United States and also from the East and West Coasts of the country is very important for Austin's success. They came by the thousands and changed Austin forever. Such an opportunity will not be available to countries with closed immigration policies, such as Japan.

The Austin Model of the 1980s and 1990s may no longer be applicable to the Digital Economy of 2006 and beyond. The importance and the reality of industrial clusters in the digital global economy is changing continuously. Everyone must be ready to adjust to accommodate these changes in order for clusters to be sustainable.

Acknowledgements

The author wishes to acknowledge Dr. Akifumi Kuchiki and Mr. Kentaro Yoshida of IDE-JETRO of Japan for providing the opportunity to conduct this research. Also, the author expresses his appreciations to Ms. Jessica Wozniak and Prof. David Eaton of the Lyndon B. Johnson School of Public Affairs and to Dr. David Gibson of the IC^2 Institute, University of Texas, for their valuable advice and contributions to this research.

Notes

1. Scott, Bruce R. and Srinivas Sunder, "Austin, Texas: Building a High-Tech Economy," Boston: Harvard Business School Publishing, 1998, p. 18.
2. Akifumi Kuchiki, Sun Jian and Juan J. Palacios, *Comparison of Industrial Agglomerations Between Asia and the Other Regions*, Japan: Institute of Developing Economies (2005).
3. Annex III The Scope of the Work on a Joint Study on "Comparison of Industrial Clusters Between Asia and the Other Regions."
4. Ibid.
5. Ibid.
6. Ibid, pp. 49–50.
7. Kuchiki, Akifumi, Sun Jian and Juan J. Palacios, "Comparison of Industrial Agglomerations Between Asia and the Other Regions", Japan: Institute of Developing Economies, 2005, p. 151.
8. Smilor, Gibson, Kozmetsky, p.49.
9. In 1951 a partnership between academia and industry led to the development of a high technology industrial park on land owned by Stanford, which became known as Silicon Valley. This partnership was marked by frequent interaction between the university and business leaders through faculty consulting, industrial researchers teaching specialized courses, and an honors program in which corporate employees could earn their degrees while working full-time. The model was a resounding success. High-tech firms flocked to the region, creating an economic cluster. However, replicating the Stanford model proved to be extremely difficult.
10. Scott, Sunder, p. 2.
11. Smilor, Gibson, Kozmetsky, p. 60.
12. Scott, Sunder, p. 2.
13. Ibid.
14. Ibid.
15. Ryan Robinson, City Demographer, Department of Planning, City of Austin. January 2005. Online. Available: austin_forecast05_annual_pub.xls. Accessed: July 14, 2005.
16. Scott, Sunder, p. 19.
17. Austin Energy. Online. Available: http://www.austinenergy.com/About%20Us/Company%20Profile/powerPlants.htm. Accessed: July 12, 2005.
18. LCRA web site. Online. Available: www.lcra.org/about/overview.html. Accessed: July 13, 2005.
19. Greater Austin Chamber of Commerce, Telecommunications & Utilities Overview.

20. Ibid.
21. Ibid., pp. 6–7.
22. Permanent University Fund (PUF), Higher Education Assistance Fund (HEAF) Overview, p. 1. Online. Available: http://www.thecb.state.tx.us/reports/pdf/0434.pdf. Accessed: July 18, 2005.
23. Smilor, Gibson, Kozmetsky, p. 53.
24. Online. Available: http://www.ce.utexas.edu/dept/news/puf.html. Accessed: July 18, 2005.
25. Permanent University Fund (PUF), Higher Education Assistance Fund (HEAF) Overview.
26. Scott, Sunder, p. 10.
27. Ibid.
28. Scott, Sunder, p. 10.
29. Ibid.
30. Ibid.
31. Ibid.
32. Scott, Sunder, p. 8.
33. Ibid.
34. Wiggins, Joel and David V. Gibson. Overview of US incubators and the case of the Austin Technology Incubator, p. 56. International Journal Entrepreneurship and Innovation Management, Vol. 3., Nos. 1/2, 2003, p. 60.
35. Wiggins, Gibson, p. 59.
36. Ibid.
37. IC2 and Scott, Sunder, p. 17.
38. Ibid.
39. Ibid., p. 11.
40. The average annual incomes were calculated by multiplying the weekly wages by 50 weeks.
41. Ibid.
42. Greater Austin Chamber of Commerce.
43. Scott, Sunder, p. 12.
44. Greater Austin Chamber of Commerce.
45. Scott, Sunder, p. 13.
46. Ibid., p. 12.
47. Institute of International Education "Open Doors 2004" annual report. Online. Available: http://opendoors.iienetwork.org/?p=50137. Accessed: November 3, 2005.
48. Immigrant and Minority Entrepreneurship: The Continuous Rebirth of American Communities. Edited by John Sibley Butler and George Kozmetsky. Connecticut: Praeger, 2004, p. viii.
49. Greater Austin Chamber of Commerce.
50. Scott, Sunder, p. 11.
51. Ibid.
52. Greater Austin Chamber of Commerce.
53. Ibid.
54. "New book: Austin among most nature-friendly areas in U.S." *Austin Business Journal*, June 24, 2005. Online. Available: http://austin.bizjournals.com/austin/stories/2005/06/20/daily42.html. Accessed: June 25, 2005.
55. Scott, Sunder, p. 18.
56. Texas Almanac: State Facts and Figures - SHG Resources http://www.shgresources.com/tx/almanac/

57. http://www.cityrating.com/cityweather.asp?City=Austin
58. Smilor, Gibson, Kozmetsky, p. 57.
59. Ibid, p. 59.
60. Ibid.
61. Ibid., p. 60.
62. Ibid.
63. Scott, Sunder, p. 3.
64. Scott, Sunder, p. 3.
65. Smilor, Gibson, Kozmetsky, p. 51.
66. Scott, Sunder, p. 3.
67. Smilor, Gibson, Kozmetsky, pp. 50–51.
68. Ibid, pp. 53–55.
69. Scott, Sunder, p. 13.
70. Smilor, Gibson, Kozmetsky, p. 55.
71. Scott, Sunder, p. 13.
72. Institute of International Education "Open Doors 2004" annual report. Online. Available: http://opendoors.iienetwork.org/?p=50137. Accessed: November 3, 2005.
73. Ibid, p. 56.
74. Ibid.
75. Scott, Sunder, p. 14.
76. Ibid, p. 15.
77. Ibid.
78. IC2.
79. Scott, Sunder, p. 16.
80. Ibid., p. 15.
81. Wiggins, Gibson, pp. 57–58.
82. Ibid, p. 58.
83. Ibid., p. 60.
84. AIT Brochure and Web site. Online. Available: http://www.ic2-ati.org/. Accessed: July 19, 2005. and Wigging, Gibson, pp. 56–57.
85. Smilor, Gibson, Kozmetsky, p. 56.
86. Scott, Sunder, p. 17.
87. Ibid.
88. Ibid., pp. 17–18.
89. Ibid, p. 19.
90. Scott, Sunder, pp. 2–3.
91. Report on the microelectronics and computer technology conference. Online. Available: http://portal.acm.org/citation.cfm?id=4021.214934. Accessed: July 19, 2005. Also, Scott, Sunder, p. 3.
92. SEMATECH. Online. Available: http://www.SEMATECH.org/corporate/history.htm. Accessed: July 19, 2005.
93. Ibid., p. 63.
94. Interview Robert Cullick, Executive Manager of Corporate Communications, July 14, 2005.
95. Scott, Sunder, p. 3.
96. Scott, Sunder, p. 9.
97. Ibid.
98. Smilor, Gibson, Kozmetsky, p. 52.
99. Ibid.

100. Gibson, Rogers
101. Smilor, Gibson, Kozmetsky.
102. Ibid., pp. 49–50.
103. Smilor, Gibson, Kozmetsky, p. 64.
104. Ibid.
105. Ibid.
106. Ibid.
107. Ibid.
108. Ibid.
109. Scott, Saunder, p. 1.
110. Smilor, Gibson, Kozmetsky.
111. Scott, Saunder, p. 1.
112. Smilor, Gibson, Kozmetsky.
113. Scott, Sunder, p. 19.
114. Kuchiki, Akifumi, Sun Jian and Juan J. Palacios, Comparison of Industrial Agglomerations Between Asia and the Other Regions. Japan: Institute of Developing Economies, 2005, p. 151.
115. Scott and Sunder, p. 9.

Bibliography

Annex III. The Scope of the Work on a Joint Study on "Comparison of Industrial Clusters Between Asia and the Other Regions."

ATI Brochure and Web site. Online. Available: http://www.ic2-ati.org/. Accessed: July 19, 2005.

Austin Business Journal. "New book: Austin among most nature-friendly areas in U.S." June 24, 2005. Online. Available: http://austin.bizjournals.com/austin/stories/2005/06/20/daily42.html. Accessed: June 25, 2005.

Austin Energy. Online. Available: http://www.austinenergy.com/About%20Us/Company%20Profile/powerPlants.htm. Accessed: July 12, 2005.

Austin's Wireless Future. IC² Institute, University of Texas at Austin, January 2004.

Boisseau, C. *High Tech Dependent on Plenty of Clean Water.* April 2005. Online. Available at www.lcra.org/featurestory/hightechwater.html. Accessed: July 13, 2005.

Butler, J. S. and G. Kozmetsky (eds) *Immigrant and Minority Entrepreneurship: The Continuous Rebirth of American Communities.* Connecticut: Praeger, 2004.

City of Austin, City Demographer Ryan Robinson, Department of Planning, City of Austin. January 2005. Online. Available at austin_forecast05_annual_pub.xls. Accessed: July 14, 2005.

City Ratings. Online. Available at http://www.cityrating.com/cityweather.asp?City=Austin

Cullick, R., Executive Manager of Corporate Communications for LCRA. Interview, July 14, 2005.

Dell at a Glance. Online. Available at http://www1.us.dell.com/content/topics/global.aspx/corp/background/en/facts?c=us&l=en&s=corp&~section=003. Accessed: July 18, 2005.

Gibson, D. and E. Rogers. *R&R Collaboration on Trial.* HBR Press, 1994.

Greater Austin Chamber of Commerce. *Transportation Overview, Public Transit, Telecommunications & Utilities Overview.*

Innovation Creativity & Capital (IC²). Online. Available at : http://www.ic2.org/main.php?a=5&s=0. Accessed: July 12, 2005.

Institute of International Education "Open Doors 2004" annual report. Online. Available at: http://opendoors.iienetwork.org/?p=50137. Accessed: November 3, 2005.

Kuchiki, A., Sun Jian and Juan J. Palacios, *Comparison of Industrial Agglomerations Between Asia and the Other Regions*. Japan: Institute of Developing Economies, 2005.

Lower Colorado River Authority (LCRA) Web site. Online. Available at : www.lcra.org/about/overview.html. Accessed: July 13, 2005.

MCC. Online. Available at: http://foldoc.doc.ic.ac.uk/foldoc/foldoc.cgi?Microelectronics+and+Computer+Technology+Corporation. Accessed: July 19, 2005.

Ovetz, R. *Capitalism in West Campus: Exposing and Resisting George Kozmetsky's IC2*. Online. Available at: http://www.utwatch.org/archives/resistIC^2.html. Accessed: June 15, 2005.

Permanent University Fund (PUF) Higher Education Assistance Fund (HEAF) Overview, p. 1. Online. Available at: http://www.thecb.state.tx.us/reports/pdf/0434.pdf. Accessed: July 18, 2005.

PUF. Online. Available at: http://www.ce.utexas.edu/dept/news/puf.html. Accessed: July 18, 2005.

Report on the microelectronics and computer technology conference. Online. Available at: http://portal.acm.org/citation.cfm?id=4021.214934. Accessed: July 19, 2005.

Scott, B. R. and S. Sunder. *Austin, Texas: Building a High-Tech Economy*. Boston: Harvard Business School Publishing, 1998.

SEMATECH. Online. Available at: http://www.SEMATECH.org/corporate/history.htm. Accessed: July 19, 2005.

Smilor, R. W., D. V. Gibson and G. Kozmetsky. *Creating the Technopolis: High-Technology Development in Austin, Texas*. Journal of Business Venturing 4, p. 49. New York: Elsevier Science Publishing Co., 1988.

Texas Almanac: State Facts and Figures - SHG Resources http://www.shgresources.com/tx/almanac/

Trilogy Web site. Online. Available: http://www.trilogy.com/Sections/Careers/opportunities/. Accessed: July 19, 2005.

University of Texas at Austin Student Body Profile 2003. Online. Available at: http://www.utexas.edu/student/admissions/stuprofile/studentbody.html.Accessed: November 3, 2005.

Wiggins, J. and D. V. Gibson. *Overview of US incubators and the case of the Austin Technology Incubator*. International Journal Entrepreneurship and Innovation Management, vol. 3 No. 1/2, p. 56. 2003.

Yahoo Financial. Online. Available at: http://uk.finance.yahoo.com/q/hp?s=DELL&b=17&a=07&c=1988&e=9&d=06&f=2005&g=m.

5

Automobile Clusters in India: Evidence from Chennai and the National Capital Region[1]

Aya Okada and N. S. Siddharthan

1. Introduction

Empirical studies on agglomeration economies have primarily focused on the nature and sources of agglomeration, its impact on the performance of firms and industries, and the mechanisms that connect agglomeration to innovation and regional growth (Saxenian 1994; Breschi and Lissoni 2001; Rosenthal and Strange 2004). With regard to the sources of agglomeration economies, the classical literature, in particular Marshall's (1920) seminal work, discussed highly localized industries and their contribution to the continued growth of the town and the industry (1920: 271), identifying three key factors: increasing returns to scale, labor market pooling, and knowledge spillovers. More recent works, particularly theoretical advances in economic geography, have expanded and built on Marshall's classic arguments (Fujita et al., 1999; Rosenthal and Strange 2004). Some scholars have identified other relatively immobile resources, such as knowledge, skills, local institutions, and industrial and corporate structures and networks, as important determinants of growth-enhancing effects of agglomeration (Saxenian 1994; Breschi and Malerba 2001; Karlsson et al., 2005).

Some recent works, however, support localization economies rather than urbanization economies; they suggest that the effects of agglomeration vary across industries (Henderson 1986, 2003; Nakamura 1985; Rosenthal and Strange 2003). Others stress the important role of foreign direct investment (FDI) as a determinant of the patterns of agglomeration, particularly in the context of developing countries (He 2002). However, empirical literature on agglomeration and clusters in developing countries is still limited, though recognition is growing that such research is crucial for their regional and national economic development. Especially sparse is the literature on the causes of variations in the patterns of cluster formation and development across and within industries. As for India, apart from a few case studies, few systematic studies have focused on the spatial dimension of its industrial development, despite recent growing interest in India's growth prospect among scholars and policymakers both in and outside the country.

Therefore, the objectives of this chapter are threefold. First, we analyze the patterns of agglomeration of the Indian automobile industry (including automotive components). Second, we examine and contrast the factors that have led to different patterns of cluster development in two leading auto clusters in India – Chennai in the state of Tamil Nadu, and the National Capital Region (NCR). Finally, we analyze whether firms included in clusters perform better than those that are excluded and whether the relative importance of variables that determine the behavior of firms differs among clusters.

This study focuses on these two auto clusters in India for several reasons. First, the automobile industry, because of its large backward linkages, greatly influences the patterns of economic development in almost every country and every region that produces cars. Second, the automobile sector has grown remarkably since the 1980s, to become one of India's leading manufacturing industries (Okada 2000). Third, these two auto clusters, while operating in the same industry in the same country, exhibit very different patterns of agglomeration, and thus allow us to examine and contrast the factors that have led these clusters to grow differently.

This study employs a combination of quantitative and qualitative analyses. For the former, we present econometric analyses to examine the differential patterns of behavior between clustered and non-clustered firms, using the data set available from the *Capital Line* database.[2] Our sample consists of all the automobile component manufacturing firms listed in the *Capital Line* data set, covering the period 1998 to 2005.

In addition, we present qualitative analyses drawing on extensive interviews we conducted with managers at both assemblers and component suppliers in the Indian automobile industry as well as with representatives from the government agencies, industrial associations such as the Confederation of Indian Industries (CII), the Society of Indian Automobile Manufacturers (SIAM), and the Automotive Component Manufacturers Associations (ACMA).

This chapter is organized as follows. Section 2 surveys the literature on clusters, knowledge spillovers, and FDI to position this study in light of the literature. Section 3 analyzes the factors that have influenced the formation and growth of automobile clusters in Chennai and the NCR. Section 4 presents an econometric analysis of the differences in the conduct and performance of firms in the three main auto clusters in India – Chennai, the NCR, and the Mumbai-Pune belt (in the state of Maharashtra) – and firms that are outside the three clusters. Section 5 summarizes our findings and brings out the main lessons from the study.

2. Factors contributing to the growth of clusters

Many studies have searched for the factors that explain the sources and growth of agglomeration and industrial clusters. Scholars have identified factors such as increasing returns to scale, labor pooling, and knowledge spillovers as the main reasons for cluster formation (Breschi and Malerba 2001). Various scholars, particularly endogenous growth theorists, have focused on the role of knowledge spillovers in generating increasing returns (Romer 1986; Krugman 1991; Grossman and Helpman 1991). Agglomeration induces knowledge spillovers and thereby innovation. Indeed, some evidence shows that physical distance from the knowledge source influences R&D spillovers (Acs et al., 1994; Feldman 1994). Audretsch and Feldman (1996), however, found that even after controlling for the degree of geographic concentration in production, there is a clustering of innovative activities in industries where knowledge spillovers play a decisive role.

Other recent studies have emphasized the role of universities, educational institutions, and public laboratories in promoting knowledge spillovers and thereby encouraging cluster formation (Zucker et al., 1998; Audretsch and Lehmann 2005). Similarly, other studies stress the role of region-specific characteristics, such as a region's learning capability involving close interactions and networking among firms and other institutions within the region, in explaining regional innovative activities (Cooke 2001; Breschi and Malerba 2001; Ronde and Hussler 2005). The empirical results of Ronde and Hussler (2005) suggest a necessary condition for regional innovation: the presence of a highly qualified and skilled labor force along with good universities. Likewise, the study by Asheim and Coenen (2005) on Nordic clusters also emphasizes the supply of a highly skilled labor force and access to scientific excellence as key factors. They make an interesting distinction between clusters and regional information systems, which though strongly interrelated are yet different concepts.

With respect to the factors that promote the growth of clusters, the presence of FDI seems critical particularly in the context of developing countries. Indeed, a rich literature exists on favorable links between FDI inflows and industrial clusters. Statistical results from several studies focusing on developing economies strongly buttress the argument that foreign investors are inclined to favor locations that could minimize information costs and offer a variety of agglomeration economies (He 2002). Belderbos and Carree (2002) confirm the major impact of regions in promoting industry, and Japanese *keiretsu*-specific agglomeration benefits in China. Likewise, Tuan and Linda (2003) find that with given distance from the core, firms prefer sites with higher firm agglomeration. The pattern of agglomeration may also influence the sectoral pattern of FDI across countries or the inter-country distribution of FDI flows in a particular sector (Eaton, Lipsey and Safarian 1994).

Wei (1999) analyzes the determinants of the regional distribution of FDI within China and finds a long-term relationship between the spatial distribution of FDI and a number of regional characteristics. Provinces with higher levels of international trade, lower wage rates, more R&D manpower, higher GDP growth rates, quicker improvement in infrastructure, more rapid advances in agglomeration, more preferential policies, and closer ethnic links with overseas Chinese attract relatively more FDI. Likewise, a study on Indonesia by Syamwil et al. (2000) indicates continuous regional concentration in the core region of Java and that markets, agglomeration and infrastructure continue to be the main reasons why Japanese manufacturing industries locate in the region.

Evidence from developed economies also suggests strong linkages between FDI and agglomeration (Ford and Strange 1999). Agglomeration economies, local industry output, educational attainment, and English language ability have significantly positive effects on the location decisions of firms investing abroad, whereas wage levels, unionization, and local industry productivity all had significantly negative effects.

Head et al. (1995) argue that firms in the same industry may be drawn to the same locations because proximity generates positive externalities or agglomeration effects, and that chance events and government inducements can have a lasting influence on the geographical pattern of manufacturing. In another study Head and Ries (1996) find that "attractive" cities – those with good infrastructure and an established industrial base – are a determinant for FDI location.

Kuchiki (2004) in what he calls "a flowchart model" stresses the role of three factors in developing industrial clusters: 1) policy interventions (such as the creation of industrial zones and export processing zones); 2) local capacity building (infrastructure, institutions, and skills); and 3) the presence of anchor firms.[3] The first two factors lead anchor firms to locate in the region, and as related firms follow the anchor firms, the cluster grows. This model focuses on policy-induced processes in the development of industrial clusters, rather than autonomous processes in the conventional Marshallian-type industrial agglomerations.[4]

The studies surveyed in this section suggest that complex interactions of various factors, including idiosyncratic local characteristics, historical conditions, and industrial policy, contribute to the formation and development of industrial clusters.

3. A study of two auto clusters: Chennai and the national capital region

In this section we focus on two leading clusters in automobiles, one of the key manufacturing industries in India, namely, Chennai (the state capital of

Tamil Nadu) and the NCR, and consider the factors that have contributed to the development of these clusters.

3.1. The evolution of the Indian automobile industry

Before considering the two clusters, we briefly outline the historical evolution of auto clusters in India. India's automotive industry got its start in the mid-1950s. Before then, India imported cars and most spare parts. Large family-owned conglomerates, locally known as business houses, took the lead in importing cars and spare parts. The main importers of spare parts were the TVS group and the Amalgamations group in Madras (now renamed as Chennai), Anand and Nandas (Escorts) in Delhi, and Doshi in Bombay (now Mumbai). In 1957, the Indian Tariff Commission decided to discourage imports and encourage the manufacturing of automobiles and spare parts in India as part of India's inward-looking import substitution industrialization strategy. Consequently, the main importers began to engage in manufacturing. The Birla group (Hindustan Motors) set up a plant in Calcutta to produce the Ambassador. So did Doshi (Premier Automobiles) in Bombay, Standard Motors (Standard Herald) in Madras, and Ashok Leyland in Madras, to assemble Leyland trucks and chassis. The Tata Engineering and Locomotive Co. Ltd. (TELCO; recently renamed as Tata Motors, one of over 120 affiliated firms of the Tata group) set up in Jamshedpur in the state of Bihar to produce Tata trucks.[5] Thus, auto clusters started to emerge in Mumbai and Chennai (see Map 1).

Out of these initial ventures, only TELCO and Ashok Leyland have continued as important players. This occurred partly because the government implemented its 1963 Monopolies and Restrictive Trade Practices (MRTP) policy, which introduced an extensive licensing regime, and restricted the activities of large private business houses in order to control monopolies and the growth of private firms and to promote public-sector enterprises (Encarnation 1989; Okada 2000). Moreover, starting in the early 1970s, the government restricted the inflow of FDI, both in terms of the sectors to enter and the equity share, to promote the localization of the domestic industries and to limit FDI only to the sectors that operated in accordance with the government's priorities (Lall 1987; Encarnation 1989; Okada 2000).[6] These policies led to considerable stagnation in the automobile industry in terms of both output growth and technological development for more than two decades until the mid-1980s, when the government gradually started lifting these inward-looking and restrictive policies, and entered into a joint venture with Suzuki Motors to establish MUL.

As the Indian government introduced the new economic and industrial policies in 1991, the Indian automobile industry has experienced rapid transformations, with many new entrants forming joint ventures with foreign car manufacturers, which has drastically changed the structure of

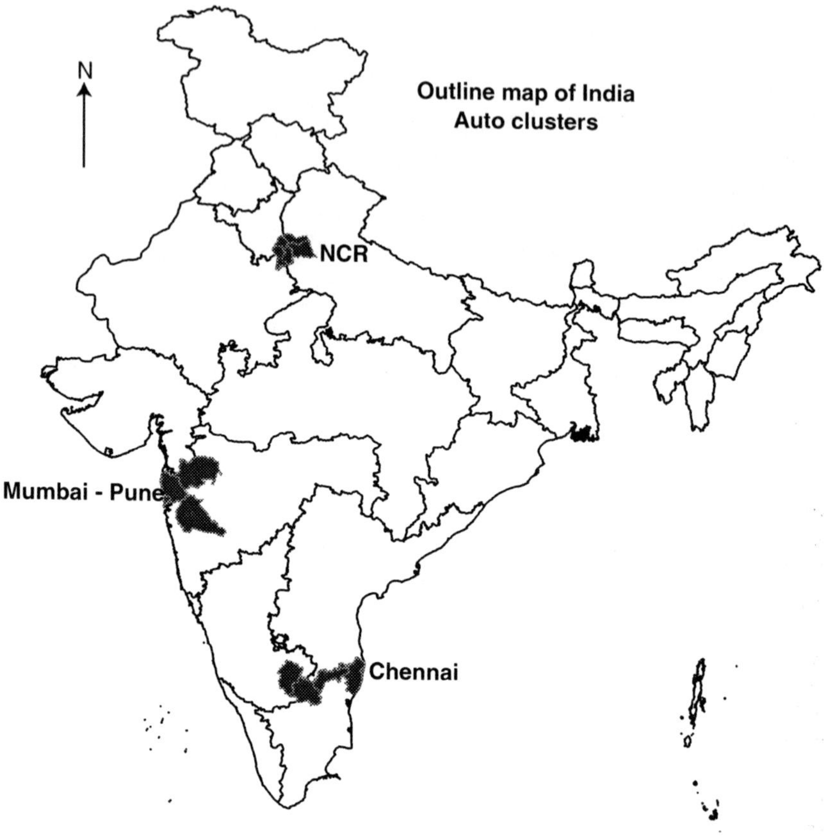

Map 1 The three prominent clusters

the automobile industry (Okada 2000). By the mid-1990s, the Indian automobile industry consisted of about two dozen assemblers of different vehicle types and the component manufacturing industry, which in turn consisted of about 350 large and medium-scale firms in the organized sector, and approximately 6,000 small firms, which are mostly in the unorganized sector (ACMA 1995).

3.2. The Chennai auto cluster

Currently, as a leading auto cluster in India, Chennai (or Tamil Nadu) accounts for 21 percent of the passenger cars, 33 percent of the commercial vehicles and 35 percent of the auto components produced in India. At present, over 100 medium and large auto companies are located in and around the Chennai cluster. Table 5.1 lists key auto firms whose plants are

Table 5.1 List of key auto firms located in Chennai cluster

Name of firm	Year of est.	Main products	FDI involvement (%) **/affiliation with Indian business houses
Assemblers			
Ashok Leyland Ltd.	1948	Commercial vehicles	FDI 50.93% (JV)
Ford India Ltd.	1995	Passenger cars	FDI (MNE subsidiary)
Hindustan Motors Ltd.	1942*	Passenger cars	
Hyundai Motor India Ltd.	1996	Passenger cars	FDI (MNE subsidiary)
Standard Motor Products of India Ltd.	1948	Passenger cars (out of business)	
Component Manufacturers			
A B I Showatech (India) Ltd.	1991		
Amalgamations Valeo Clutch Ltd.	1997	Clutch parts	(Amalgamations group)
Amalgamations Repco Ltd.	1967	Clutch products	(Amalgamations group)
Automobile Products of India Ltd.	1949	Clutch and brake assembly	
Axles India Ltd.	1981	Axiles	FDI 45.48% (JV)
Bharat Technologies Auto Components Ltd.	1992		
Brakes India Ltd.	1962	Brakes	(TVS Group)
Delphi-TVS Diesel Systems Ltd.	1969	Fuel injection equipment	
Engine Valves Ltd.	1954	Engine valves	
Exide Industries Ltd.	1947*	Auto batteries	FDI 50.90% (JV)
H S I Automotives Ltd.	1998		
Harita Seating Systems Ltd.	1986	Seats	FDI 30.58% (JV)
India Motor Parts & Accessories Ltd.	1954	Auto parts and accessories	
India Japan Lighting Ltd.	1996	Automotive lighting	
India Nippon Electricals Ltd.	1984	Flywheel Magneto, CDI unit, ignition coil	FDI 20.52% (JV)
India Pistons Ltd.	1949	Pistons and piston rings	(Amalgamations group)
Indiapistons-Repco Ltd.	1963	Pistons	(Amalgamations group).

Continued

Table 5.1 Continued

Name of firm	Year of est.	Main products	FDI involvement (%) **/ affiliation with Indian business houses
IP Rings Ltd.	1991	Piston rings	
JBM Sung Woo Ltd.	1998	Steel metal stamping and assemblers	
JKM Dae Rim Automotive Ltd.	1997	Exhaust systems, axle assembly, water pump components	
Jonas Woodhead & Sons (India) Ltd.	1963	Leaf springs	
Lakshmi Auto Components Ltd.	1986	Plastic and rubber moulded components	
Lucas – TVS Ltd.	1961	Auto electrical products including starters, motors and ignition products	(TVS group)
Lucas Indian Services Ltd.	1986	Auto electrical products ignition coils, rotors	(TVS group)
Mando Brake Systems India Ltd.	1997	Brake systems	
Motherson Sumi Systems Ltd,	1986	Auto electrical wiring and cords	FDI **34.7% (JV)**
P H C Manufacturing Ltd.	1997		
Pentadaewha Auto Parts Ltd.	1997		
PRICOL Ltd.	1972	Engine tools and castings, dashboard instrument, oil pumps	**FDI 12.50% (JV)**
Rane (Madras) Ltd.	1960	Steering, ball joints, axial joints, suspension joints	(Rane group)
Rane Brake Linings Ltd.	1964	Brake lingings	**FDI 10%** (Rane group)
Rane Engine Valves Ltd.	1959	Engine valves	(Rane group)
Rane Holdings Ltd.	1929	Main holding co. of Rane group	(Rane group)
Rane N S K Steering Systems Ltd	1995	Steering	(Rane group)
Rane T R W Steering Systems Ltd.	1987	Steering systems	(Rane group)
Renowned Auto Products Mfrs Ltd.	1997	Shock absorbers	
Roots Industries Ltd.	1990	Auto electricals	
Shardlow India Ltd.	1960		
Simpson & Company Ltd.	1925	Diesel engines for automobiles and tractors	(Amalgamations group)

Continued

Table 5.1 Continued

Name of firm	**Year of est.**	**Main products**	**FDI involvement (%) ** / affiliation with Indian business houses**
Sona Koyo Steering Systems Ltd.	1984*	Steering systems	FDI 20.47% (JV)
Standard Batteries Ltd.	1945	Batteries	
Sundaram Auto Components Ltd.	1996	Plastic and rubber moulded components	(TVS group)
Sundaram Brake Linings Ltd.	1974	Brake linings	(TVS group)
Sundaram Clayton Ltd	1962	Air brakes, vaccum brakes, and aluminium castings	FDI 39.17% (JV) (TVS Group)
Sundaram Industries Ltd.	1942	Engine mounts, suspension bushes, CVJ boots and body building	(TVS Group)
Super Auto Forge Ltd.	1974	Base balls – iron and steel.	
Super Shock Absorbers Ltd.	1983	Shock absorbers	
Talbros Automotive Components Ltd.	1956	Gaskets	FDI 30% (JV)
Tractors & Farm Equipment Ltd.	1960	Tractors and farm equipment	(Amalgamations group)
TVS Autolec Ltd .	1980	Pump assemblies and engine components	(TVS group)
TVS Motor Company Ltd.	1982	Motor cycles and two wheelers	(TVS group)
Ucal Fuel Systems Ltd.	1985	Fuel systems	FDI 23.32% (JV)
Wheels India Ltd.	1960	Auto wheels	FDI 35.91% (JV) (TVS group)

Notes: Firms that have plants in the Chennai cluster are listed.
* The year of the establishment of the firm, not of its Chennai plant. The date of the establishment of the firm's Chennai plant is not known.
** FDI is the percentage of foreign promoter's equity in the total equity as of 2005/06. Based on this we have classified the firms as joint ventures (JV). Many firms have foreign technical collaboration but not equity participation. Some of the older firms had foreign equity in the initial years but now do not have foreign equity.

Source: *Capital Line* database.

located in the Chennai cluster. In the auto component industry, Tamil Nadu has a more than 50 percent share in the production of many key parts, including engines, inlet and exhaust valves, fuel pumps, starter motors, camshafts, oil seals, wiper motors, and air brake assembly and engines. It produces between 30 percent and 50 percent of parts such as voltage regulators, steering gears, wheel rims, electric horns, and dashboard instruments, and between 10 percent and 30 percent of many more items, including crankshafts, radiators, clutch plates, shock absorbers, tyres and automotive seats. Most of these firms have ISO certification and about a quarter of them have QS certification. Chennai has emerged as one of India's leading auto clusters because of several historical, political and economic factors, which we discuss below.

3.2.1. The emergence of the Chennai auto cluster

First, when the Tariff Commission decided to discourage imports and favor the domestic production of automobiles and spare parts, several Tamil Nadu-based firms like TVS and Standard Motors were already at the forefront. For example, the TVS group set up plants in Madras (Chennai) and developed an important industrial enclave in Padi on the outskirts of the city. The emergence of Madras, Bombay, and Calcutta as important auto clusters until the early 1960s is partly because these cities had important seaports. Clearly, proximity to a seaport was an important consideration for the formation of the auto clusters in the earlier years, because the industry (including the component sector) depended on imports until the early 1960s.

Second, during the late 1950s and the early 1960s, the government of Tamil Nadu (earlier called Madras State) actively supported and promoted these firms. Particularly involved were its Chief Minister K. Kamaraj, the Industry Minister R. Venkataraman (who later became a member of the Indian Planning Commission and the President of India) and T. T. Krishnamachari.[7] These political leaders were instrumental in giving industrial licenses to firms to set up plants for manufacturing trucks and other heavy vehicles in and around Chennai.

Third, several Tamil Nadu-based industrialists, such as those of the TVS group, MRF, Ashok Leyland, Standard Motors, and the Rane group, played notable roles in forming the auto cluster in Chennai.[8] For example, Mammen Mappillai started out with a toy balloon plant in a small shed in Tiruvottiyur, a Madras suburb, in 1946, and became one of the largest tyre manufacturers in India. In 1952, he started manufacturing tread rubber, and his firm gradually blossomed into MRF as he entered into technical collaboration with Mansfield Tire and Rubber Company, US, in 1961. By 1967, MRF had started exporting tyres to the U.S.[9]

3.2.2. Capacity building: infrastructure, human resources and institutions

Several additional factors explain the subsequent development of the Chennai cluster after its initial formation in the 1950s. One important factor that attracts FDI and domestic firms in the auto and information technology (IT) industries in Chennai is the adequate infrastructure – both physical (particularly reliable telecommunication network), and human resources (a steady supply of skilled workforce), as well as government intervention in terms of investments in technology parks including IT and auto parks. With regard to human resources, Tamil Nadu produces the largest number of engineering graduates in the country (Ramachandran and Goebel 2002). Chennai is home to several engineering colleges, including the Indian Institute of Technology (IIT), the Guindy engineering college, and the A. C. College of Technology. Some regional engineering colleges and several private engineering colleges were also established in Tamil Nadu. As industrial estates were set up, several technical institutes were simultaneously set up to train technicians and mechanics. Moreover, the state government introduced a mid-day meal program in primary and secondary schools to improve attendance in the schools and reduce drop-out rates. Thus, from primary schools to highly sophisticated engineering institutes, Tamil Nadu already had the education institutions in place as early as the late 1950s.

Furthermore, during the 1980s, several new engineering colleges were set up both by the government and the private sector. Lower-level technical institutions such as the government-run Industrial Training Institutes (ITI) to train technicians proliferated, and consequently Tamil Nadu emerged as the state that produced the largest number of technical graduates. The state government has also involved the private sector in training workers and professionals. At the same time, the mid-day meal program in primary and secondary schools was strengthened and the scope widened. As a result, as discussed earlier, 99 percent of Tamil Nadu children in the age group of 11–14 attend schools; this is the highest percentage in India.

With regard to infrastructure, Chennai has an international airport and two sea ports, the second one recently constructed at distance of 25 kilometres from the city. In addition to the two Chennai ports, Tamil Nadu has an all-weather port at Tuticorin, as well as two intermediate and six minor ports.

Moreover, the state government gives a number of financial incentives like capital subsidies, power tariff concessions, sales tax waiver and other schemes to firms located at the industrial estates created by the state government. Moreover, four major state agencies are actively involved in soliciting and facilitating FDI in the state: the Electronic Corporation of Tamil Nadu Ltd; the State Industrial Promotion Council of Tamil Nadu Ltd.; Tamil Nadu Industrial Guidance and Export Promotion Bureau; and the Electronics Test and Development Center.

The human and physical infrastructure provided by the government and the presence of large component manufacturing firms attracted global firms like Ford, Hyundai and Mitsubishi to set up plants in Chennai. The arrival of these firms had a major impact on the Chennai auto cluster resulting in a radical restructuring of the industry. Up until the early 1990s, Tamil Nadu was mainly producing components for the Indian market. Chennai's main auto component suppliers, like the TVS and Amalgamations groups, made strategic use of the government's earlier policies like subsidized access to overseas technology and support for participation in international trade fairs: now they have become globally competitive, able to face the current WTO regime (Tewari 2003).

3.2.3. Lead firms in the Chennai cluster

Several Chennai-based firms, which are mostly Indian conglomerates, have played a critical role as lead firms in the development of the Chennai auto cluster, including the TVS Group, the Rane Group, and Ashok Leyland Ltd. The TVS Group started as a transport company in 1911 and now has over 29 companies as India's leading suppliers of automotive components, many with FDI involvement.[10] For example, a group firm, Wheels India Ltd., was set up as a joint venture between TVS and Dunlop, UK, in 1960; located in Padi on the outskirts of Chennai: it produces wheels for a wide range of products, such as cars, tractors, commercial vehicles, jeeps, tractors, earth moving equipment, and defense vehicles. It supplies parts to all the vehicle manufacturers in India. Another group firm, Lucas-TVS, a joint venture between the Lucas Variety group, UK and the TVS group, was established in 1961 and is a leading manufacturer of auto electrical products and diesel fuel injection equipment in India. It covers all segments of the auto industry: passenger cars, jeeps, light and heavy commercial vehicles, industrial engines, tractors, and two wheelers. Its products include starter motors, alternators, regulators, fan motors, ignition systems, and horns.

Other group firms such as Brakes India, Sundaram Clayton Ltd., and Sundaram Fastners Ltd., were all established in the 1960s, as joint ventures with British firms, and are all located in Padi. Brakes India, which manufactures brake equipment, caters to over 60 percent of the domestic OEM market and exports to over 35 countries. Sundaram Clayton Ltd. was the first firm to manufacture brake systems in India and has been the market leader since its inception. It exports to the U.S., the U.K., Egypt, Australia, Sri Lanka, the Middle East and East Asian countries. Sundaram Fasteners Ltd. has emerged as the largest manufacturer of high tensile fasteners in India and is the key supplier of radiator caps to General Motors (GM)'s plants in the U.S. It manufactures standard fasteners like hex head bolts, screws and nuts, and special fasteners like cylinder head bolts and nuts, and main bearing cap bolts. Similarly, other Chennai-based group firms established in the 1980s and the early 1990s, such as Turbo Energy Ltd.,[11] India Nippon Electricals

Ltd., and Sundaram Dynacast Ltd, also have technical and financial collaboration with foreign component manufacturers in Germany, Japan, and the U.S., producing key components such as turbo-chargers, electronic ignition systems, and precision engineering components.

Likewise, many firms in the Rane Group manufacture automotive components in Chennai. Its flagship firm, Rane Engine Valves Ltd., was established in 1959, manufacturing engine valves, valve guides, crank shafts for compressors and clutch boosters, which are exported widely. Its other main group firms, including Rane Brake Linings, Rane Madras, Rane NSK Steering Systems, and Rane TRW Steering Systems, produce a wide range of products. Clearly, these industrial houses (conglomerates) have played a critical role in developing Chennai's auto cluster.

Ashok Leyland Ltd. (ALL), established in 1948 as Ashok Motors, the second-largest manufacturer of medium/heavy commercial vehicles in India today, also initially started to assemble Austin car parts in India. In 1955, when it entered into an agreement with Leyland Motors, U.K., to manufacture Leyland vehicles, it changed its name to Ashok Leyland.[12] ALL's manufacturing plants are located at Ennore (Chennai), Ambattur (Chennai), and Hosur (Tamil Nadu). Recently it has also established plants outside Tamil Nadu – in Bhandara (Maharashtra), Hyderabad (Andhara Pradesh) and Alwar (Rajastan). During 2003/04 it produced 12,996 commercial vehicles in the medium and heavy category and exported 1,604 vehicles.

Interestingly, therefore, except for ALL, the lead firms in the Chennai auto cluster are, rather than assemblers, all component manufacturers that were established in the 1960s, long before India's motorization started. This makes the Chennai cluster unique compared to other auto clusters in India or elsewhere.

3.2.4. Industrial estates and small firms

The Chennai cluster also includes small industries. The entrepreneurial skills in many small firms have contributed to the success of the auto cluster (Sridhar 2002). Chennai has several industrial estates, providing factory space at relatively low rents and other facilities to firms operating there. The Guindy industrial estate – the largest in Chennai – was established by the state government in 1958 on a 100-acre plot. The estate provides technical services like mechanical, metallurgical and chemical testing laboratories, tool rooms, forging and a heat treatment shop and wire drawing unit. It also has a library and technical information section. Chennai also houses other industrial estates at Ambattur, Arumbakkam, Villivakkam, Kodungaiyur, Madhavaram and Perambur. The smaller firms are also beginning to globalize and export to large MNEs.

Two factors have recently facilitated the globalization process for these small firms in Chennai's industrial estates. First, the presence of a large number of IT firms in Chennai has enabled them to take advantage of

Business to Business (B2B) Commerce. Second, because the US government decided to discourage or even prevent the establishment of forging and casting firms, several MNEs have been outsourcing these activities to overseas firms, and Chennai has benefited from this emerging practice.

3.2.5. The recent growth of information technology industry and the auto cluster

Furthermore, in the 1990s onwards, the growth of the IT industry in Tamil Nadu has complemented that of the auto components sector (Tewari 2003). The Chennai auto cluster has increasingly moved into the production of parts with IT-enabled systems for both OEMs and their suppliers. Further, the rise of IT capabilities among local firms in Tamil Nadu has made it easier for non-proximate regional suppliers in the auto components sector to work closely with distant customers and their multi-locational networks (Tewari 2003). Thus, Tamil Nadu firms developed an advantage in bidding for export contracts for small, standard parts on-line and receiving requests for quotes (RFQs) from global players like GM and Ford.

The successful performance of the Chennai components industry did not result automatically from the neo-liberal de-regulation of the economy since 1991 and the arrival of MNEs in the auto sector. It occurred due to the way the government handled the de-regulation of the auto sector: the industry's current success is mainly due to the pace and the sequencing of the government's liberalization of the sector, which was highly graded and strategic (Tewari 2003).

3.2.6. The arrival of global assemblers in the 1990s

After the government de-licensed it in 1993, the automobile industry witnessed rapid transformations as many global players entered India in the 1990s, making the domestic market increasingly competitive. In 1996 and 1999, the arrival of Ford, Hyundai and Hindustan Motors Ltd. (HML) (via a licensing agreement with Mitsubishi Motors)[13] further transformed the Chennai auto cluster, as they have established passenger car production and assembly operations in Chennai. They invested $1.5 billion in Tamil Nadu (Tewari 2003) and have established a combined capacity of 230,000 cars per year: Hyundai has invested about $1 billion, Ford about $400 million and HM-Mitsubishi about $150 million. Hyundai's wholly-owned subsidiary in Chennai has quickly emerged as the second most important car manufacturer after Maruti Udyog Ltd. (MUL). From its very arrival in Tamil Nadu, Ford has been concentrating on building its global platform apart from targeting the Indian market. Consequently, Ford and Hyundai are very active in the exports market. Thus, in 2003, Ford exported 24,000 cars and Hyundai 30,000. The arrival of these MNEs has clearly boosted the components sector in the Chennai cluster, partly because these firms were required to increase their local content.[14] Indeed, in the late 1990s, both Hyundai and Ford urged their suppliers to locate in proximity (Gulyani 2001).

In 1998, Hyundai set up in Chennai a 100 percent subsidiary firm (its largest investment outside South Korea). It initially brought in about 14 South Korean component suppliers to provide parts not available in Chennai. Since then, these Korean component manufacturers have been sourcing materials and parts from small firms in industrial estates located in and around Chennai. Currently Hyundai buys about 85 percent of its components from within India, with about 50 percent from Tamil Nadu itself. Hyundai has about 70 major component suppliers; of these, only 14 are Korean joint ventures and the rest are mainly based in Tamil Nadu.

Hyundai recently announced a plan to build a second assembly plant in Chennai next to its current facility. The new plant will have an annual capacity of 150,000 units and will be constructed on a 74.2 million square foot site. This will increase Hyundai's manufacturing capacity in Chennai to 400,000 vehicles. In addition to serving the Indian market, it will cater to the export markets in Europe, Latin America and the Middle East. During 2003/04, Hyundai produced 170,942 cars: 135,008 in the compact car segment, 34,698 in the mid-size segment, and 1,236 in the premium class segment. Hyundai is the market leader in the mid-size segment, followed by Tata Motors, which produces 28,107 vehicles. In 2003/04, Hyundai exported 35,752 compact cars and 6,363 mid-size cars.

Ford imports some of its key parts but claims 75 percent local content. Local suppliers have joint ventures with US firms or have entered into technology licensing with US firms. Within two years after it began operations in Chennai, Ford opted for global sourcing of components from Tamil Nadu. It also had short-listed global suppliers from Tamil Nadu. During 2003/04, Ford sold 45,035 cars in India (44,881 mid-size and 154 premium), and exported 25,000 mid-size cars.

In contrast to Ford and Hyundai, HM-Mitsubishi depends mainly on imported components from Japan; its local sourcing is only about 30 percent. The import intensity of HM-Mitsubishi is attributed to its very low volume and concentration on the luxury segment where price competition does not prevail (Tewari 2003). The other two producers have a presence in all the segments and recognize the importance of volume: hence they prefer suppliers in Chennai and Tamil Nadu. Thus, agglomeration of the automotive component industry in Chennai, mainly led by Indian business houses with foreign collaboration, preceded the entry of large assembler firms to the cluster. Only in the last several years have foreign auto makers become the lead firms.

In sum, several factors were critical in the formation and development of the Chennai auto cluster. First, government intervention, at both the central and state levels, played a crucial role in the emergence and later development of the Chennai cluster. In the late 1950s and early 1960s, Tamil Nadu had very dynamic political leaders who were instrumental in bringing auto component firms to Chennai. The state government also established many industrial estates to promote small firms and ancillary units, many producing auto components.

Second, the presence of well-developed infrastructure, particularly access to a seaport, airport and other infrastructure facilities, encouraged entrepreneurs to opt for Tamil Nadu, and in particular Chennai. Third, the presence in Tamil Nadu of well-established Indian family-owned business houses like the TVS group, the Rane group, and the Chettiars (who are mainly bankers and business men) further helped. Fourth, Chennai and other cities in Tamil Nadu enjoy a secure supply of highly skilled workers because of several engineering colleges, including India's prestigious Indian Institute of Technology (IIT).

Fifth, small firms and ancillary units played a very important role in the development of the Chennai cluster. The TVS, Rane and the Amalgamations groups set up ancillary and component units in the late 1950s and early 1960s and were instrumental in developing industrial clusters in North Chennai. In particular, the TVS group developed the Padi enclave on the outskirts of the city. These auto component manufacturing firms from inception targeted the all-India market, not only the limited Chennai market. The scale of operations of the two auto assemblers – Standard Motors (now out of business) and Ashok Leyland – was not large enough to sustain the component manufacturers. Having established themselves firmly and having developed industrial enclaves, leading component manufacturers and the TVS group in particular, kept expanding their activities by continuously starting new ventures and adding new product lines to their existing ventures. Several small firms set up plants in the industrial estates to supply materials and smaller components to these groups. The presence of these major component-producing groups and small suppliers located in the industrial estates in and around Chennai encouraged the entry of large global players like Ford, Hyundai and Mitsubishi in the late 1990s.

Finally, the emergence of the IT industry in Chennai and other cities in Tamil Nadu, and the rapid development of internet infrastructure, have helped small and medium firms to globalize and take advantage of B2B commerce. In this respect, the patterns of development of the Chennai cluster differ considerably from those of clusters elsewhere – for example, Viet Nam and China (Kuchiki 2004) – where large assembling firms lead the process of setting up and developing ancillaries. Interestingly, in the Chennai cluster, the process went in the opposite direction.

3.3. National capital region (NCR) cluster

The NCR auto cluster is located in Haryana, Delhi and some districts of Uttar Pradesh adjacent to Delhi. Maruti Udyog Ltd. (MUL) played a leading role in its development. Thus, this cluster, by and large, follows the traditional pattern of auto clusters led by assemblers that served as lead firms.

3.3.1. The growth of the passenger car segment and the emergence of an industrial leader

MUL started in 1982 as a joint venture between the Indian government and the Japanese automaker Suzuki Motor Corporation.[15] MUL set up its first plant in Gurgaon, a primarily agricultural area gradually transforming into a newly developing industrial town in Haryana adjacent to Delhi, as a green-field plant. A close copy of Suzuki's Kosai plant in Japan, it was the first modern assembly plant in India in terms of plant layout, equipment, the organization of production and the operating principle. It set up its second plant in Gurgaon in 1992, and its third plant in NOIDA (New Okhla Industrial Development Authority), also within the NCR, in 1999. Having begun production in 1983, it has since emerged as the largest car manufacturer in India by initially focusing on the small car segment, which had been virtually untapped in the Indian market until MUL's entry. MUL's cars were 21 percent cheaper than the lowest-priced existing passenger car produced by domestic manufacturers, yet offered much higher quality, more safety features and greater fuel efficiency (Humphrey et al., 1998; Okada 2000). In response to the increased variety in consumer tastes, in the early 1990s the firm also diversified its product range, introducing new middle-sized passenger cars.

As Figure 5.1 indicates, the growth in the passenger car segment of the Indian auto industry has been largely driven by MUL. In 1996/97, it produced 336,811 passenger cars, accounting for 79.6 percent of the Indian passenger car market.[16] In 2003, it built the foundry plant of Suzuki Metal India Ltd, a joint venture between Suzuki and MUL. Interestingly, even after the entry of many foreign car manufacturers to India since the mid-1990s, which intensified the competition within the Indian market, MUL continued to remain dominant. By 2003/04, MUL increased its production to 408,911 cars, of which 216,163 were compact and 14,384 mid-sized; it does not produce cars in the premium and executive segments. It exported 39,454 compact cars and 314 mid-size cars.

In addition to MUL, in the 1990s, other foreign auto manufacturers such as Daewoo Motors India Ltd. (a joint venture with Daewoo Corporation, Korea) and Honda Siel Cars India Ltd. (a joint venture with Honda Corporation, Japan) started their operations in the NCR. However, their contributions to the NCR cluster are too modest, as Daewoo ceased its brief operation after its parent in Korea failed, and Honda Siel, which started its production only in 2000/01, still produces a small volume: in 2003/04, 17,953 mid-size cars (of which it exported 27), and 2,031 premium cars. Thus, the NCR auto cluster's development has been mainly driven by MUL's remarkable growth (as shown in Figure 5.1).

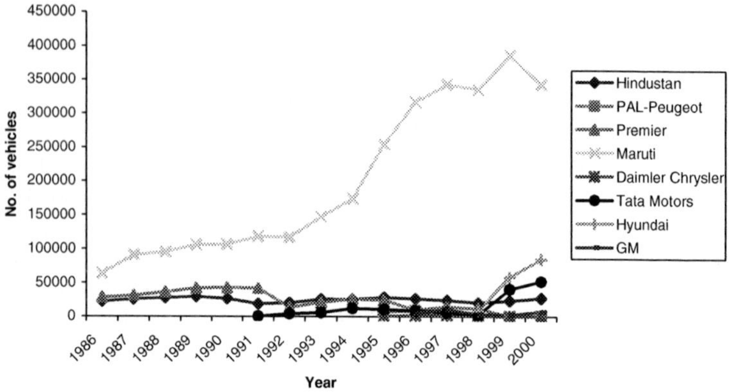

Figure 5.1 Trends in passenger car production in India by manufacturer
Sources: AIAM, 1997; SIAM, 2002.

3.3.2. MUL's supplier development strategy

MUL's experience is unique in India particularly because of its system of procurements and enormous efforts to develop first-tier suppliers in its proximity and upgrade their capabilities. MUL's supplier development was a key factor in reducing production costs, and thus a key factor in the firm's remarkable growth (see Figure 5.1). It also became a successful model for transferring know-how of the non-*keiretsu* production system (Hattori 1996). And, given its dominance in the domestic car market, and a high rate of local content (96 percent for the "Maruti 800" and over 75 percent for other models) (Okada 2000), the firm had a tremendous impact on the development of the automotive component industry as a whole, and in particular, in the NCR auto cluster.

Both MUL and Suzuki hold equities in about a dozen MUL first-tier suppliers of key components to transfer technology and develop long-term relationships. Several key suppliers (such as Bharat Seats) of important bulky items like seats, fuel tanks, bumpers and instrument panels, are located in the same complex as MUL's main plant in Gurgaon, allowing MUL to gain greater control over their operation and performance standards. Suzuki invested about 15 percent each in the equities of Bharat Seats, Macino Plastics, and Subros (a car air-conditioner manufacturer with collaboration with Denso). MUL also has 10 to 15 percent equity shares in Macino Plastics, Asahi Safety Glass and Sona Steering, about 24 percent equity in Mark Auto Industry, and 31 percent in Jay Bharat. Moreover, MUL and Suzuki arranged a joint venture between local suppliers and Suzuki's suppliers in Japan. For example, Bharat Seats has financial and technological ties with Howa Industry, which supplies seats to Suzuki in Japan; Similarly, Asahi India has ties with Asahi Glass; Sona Steering with Koyo Precision Industry; and Subros with Denso. Likewise, Motherson's was encouraged to collaborate with Sumitomo Denso to form a joint venture, Motherson Sumi, and JVC did the same with Daikin Industry.[17] Furthermore, MUL has also made enormous efforts to upgrade the capabilities of its suppliers (Okada 2004).

Table 5.2 shows the geographical distribution of MUL's first-tier suppliers as of 1997. As the table indicates, about 60 percent of these suppliers are located within the NCR cluster, i.e., within an hour's drive from MUL's plant. Apart from the key suppliers housed in MUL's complex, many other component manufacturers are located in the NCR, producing such items as lamps, mirrors, front and rear windows, clutches, shock-absorbers, and wire-harnesses. As of 1997, 58 out of the 404 MUL first-tier suppliers depended on MUL for more than 90% (in many cases 100 percent) of their sales.[18] Also, one third of the 404 MUL's first-tier suppliers were established mostly within the NCR, after MUL started its operation in 1983.[19] With MUL's remarkable growth (see Figure 5.1), which led to the rapid growth of the component industry, many second-tier suppliers have also proliferated in the same city where their customers – MUL's first-tier suppliers – were located. In other words, the NCR auto cluster emerged rapidly as MUL grew, mainly through MUL's development of local suppliers.

Table 5.2 Geographic distribution of MUL's first-tier suppliers

Geographical distribution		**# of firms**	**% distribution**	**Firm size**		
City	**State**			**Large**	**Medium**	**Small**
Faridabad*	Haryana	77	19.1	12	36	29
New Delhi*	Delhi	71	17.6	29	22	20
Gurgaon*	Haryana	63	15.6	17	19	27
Chennai	Tamil Nadu	28	6.9	17	4	7
Mumbai	Maharashtra	23	5.7	11	4	8
NOIDA*	U.P.	16	4.0	5	2	9
Pune	Maharashtra	16	4.0	10	6	0
Bangalore	Karnataka	15	3.7	7	7	1
Coimbatore	Tamil Nadu	8	2.0	4	3	1
Old Delhi*	Delhi	7	1.7	3	2	2
Ludhiana	Punjab	6	1.5	2	4	0
Ghaziabad*	U.P.	6	1.5	1	2	3
Calcutta	West Bengal	5	1.2	4	1	0
Other Locations		63	15.6	32	23	8
Total		404	100	154	135	115

Notes: * refers to locations within the NCR. The data are based on MUL's supplier database as of January 1997.

Source: MUL's internal documents (1997).

Several factors explain MUL's massive investment in its elaborate program of vendor development, involving stable and close supplier relations with its first-tier suppliers, equity participation in key suppliers, and promotion of technical collaboration between its suppliers and Suzuki's suppliers in Japan. First, the government's phased manufacturing program (PMP) mandated foreign firms to promote localization. Suzuki's MOU with its joint-venture partner, the Indian government, included its commitment to achieve 50 percent local content within the first three years, and 70 percent by the fifth year (Okada 2000). MUL's initial focus on domestic markets rather than on exports, allowed it to compromise on the quality of the component products produced by local suppliers, which it could not have afforded if it were exporting its products.

Second, the appreciation of the yen in the early 1980s, along with the high customs duty – 110 percent until 1991 – imposed on complete knock-down (CKD) parts, made imported components from Japan extremely expensive. MUL knew that if it used CKD imports, it could not compete with other domestic producers.[20] Faced with the dilemma between the poor quality of locally-produced components on the one hand and the need to increase local content on the other, however, MUL had no choice but to invest heavily in developing the capabilities of its suppliers (Okada 2000). This development of suppliers has been particularly important for MUL, given its high reliance on outsourcing which accounts for 80 percent of the value of a car, even higher than the level of outsourcing in Japan (70 percent).[21]

Third, MUL was the first firm to introduce a partial 'just in time' (JIT) system in India. This required MUL to source the dependable quality of component products, so as to avoid the need for detailed on-site inspection and quality testing by MUL after their delivery; it also required suppliers to be located near the car assembly plant, to facilitate frequent delivery and on-time delivery. However, for most parts, where the units were not located near the MUL complex, MUL had to hold a week's stock at any time (Hattori 1996). Also, MUL has a double sourcing strategy, because the generally poor infrastructure (roads, electricity, and telecommunication) often hampers on-time delivery, making it costly to rely on a single source for each component, and therefore making it difficult to fully adopt the JIT principle.[22]

Fourth, on a related point, while MUL has also sourced parts from a few large suppliers from the southern states like Tamil Nadu, these suppliers, particularly those of bulky and heavy components, were motivated to establish new plants close to their customers. This is partly because of the increased awareness among first-tier suppliers of the importance of on-time delivery through the adoption of some elements of the JIT inventory system that MUL introduced, and partly because of their interest in reducing transport costs, including the payment of a state tax called *octroi* which each state government levies on consignments each time they cross a state border.

Moreover, local suppliers in proximity could reduce their utility costs by relying on power available from MUL's in-house power plants.

Thus, given the intensified competition among component manufacturers, firms outside the NCR, such as Chennai-based Lucas-TVS, have also set up plants near MUL's plant. Indeed, with the introduction of the JIT concept, which aims to reduce assemblers' inventory costs, some first-tier suppliers with multiple customers started setting up new plants near their customers in different regions to cater to each of them. While this helps assemblers reduce their inventory costs, however, it also constrains the suppliers, as they have to manage and finance simultaneous expansions in widely dispersed locations (Humphrey et al. 1998).

Fifth, many local small firms that could serve as ancillaries had already existed prior to MUL's entry, although their technological levels were not compatible with MUL's standards. The government policy since the 1960s aimed to protect and promote the Small Scale Industries (SSI), by providing various incentives; the allocation of plots in industrial estates at subsidized costs, electricity, and telephone connection, and by reserving many auto components to be produced only by the SSI sector (Okada 2000). This reservation policy forced auto manufacturers to buy rather than make these items.[23] MUL's growth has also encouraged many small-scale entrepreneurs to start businesses in nearby locations, taking advantage of the incentive measures provided to the SSI firms. Thus, about 60 percent of MUL's 404 first-tier suppliers are small and medium-sized enterprises (see Table 5.5).

Sixth, the policy reforms introduced since 1991 played an important role in restructuring the Indian auto industry and changing the economic environment in which auto firms operated, in terms of reduction in import tariffs, intensified competition, increased inflow of FDI, and increased export opportunities not only for MUL but also for component manufacturers.

Finally, MUL's close collaboration with the government has facilitated the development of its local supplier base. The central government, the firm's joint venture partner, has protected and promoted MUL through various favorable policy measures and concessions. Moreover, while being a joint venture partner, the central government did not politically interfere with MUL's functioning, which was unusual in India. Likewise, the state government of Haryana also supported MUL in many ways. For example, in 1997, MUL set up a new supplier park near its main plant in Gurgaon as a joint venture with the Haryana State Industrial Development Corporation (HSIDC) as part of the industrial model township developed by HSIDC. It was intended to house MUL's 65 first-tier suppliers which produce essential and critical components for MUL cars, and to ensure unhindered supply of these components. HSIDC shared with MUL the cost of the development of this industrial park, at nearly Rs. 100 crore (or approx. US $30 million). A tripartite plant-level committee comprising representatives from MUL, the government, and HSIDC oversaw the planning and implementation of this project.

MUL has maintained a long-term close relationship with suppliers, based on reciprocal interactions, with a greater emphasis on quality and on-time delivery, and also provided them with designs and drawings. MUL's heavy investment in supplier development was rather unusual in India, because prior to MUL's entry, during the inward-looking trade and industrial regime until the mid-1980s, a small production volume and absence of competition gave auto assemblers few incentives to strengthen the capabilities of their suppliers. MUL not only sent its shop-floor employees to Japan for training but also encouraged their suppliers to follow its example.[24] These practices have also contributed to the development of the NCR cluster.

4. Differential behavior of cluster and non-cluster firms

The literature surveyed in Section 2 indicates that firms located in an industrial cluster enjoy several advantages like external economies, sharing of knowledge and information, technological environment, availability of the skilled labor force, infrastructure and other environmental supports provided by the government and its agencies. Because the cluster provides these advantages, the performance and conduct of the firms located in the cluster could differ from those located outside a cluster. In the automobile sector, while many firms are located in the three clusters of Chennai, the NCR, and the Mumbai – Pune, some are also located outside these clusters. We expect firms located in these three clusters to behave differently from those located outside these clusters. We also expect inter-cluster differences in the performance of the firms inside and outside the clusters. Profit margins, productivity and exports are considered as performance variables of the firms while advertisement intensity, FDI, inventories, royalty and technical fee payments, import of materials and components and other foreign payments are taken as representative of conduct variables. The conduct variables mainly represent international orientation, technology and product differentiation variables.

From the literature survey we have identified variables that could discriminate between the firms in the various clusters and other firms located outside them. We cover only auto component firms and not the leading auto assemblers. We classify firms into four groups according to their locations: (a) Chennai (including firms in surrounding areas in Tamil Nadu); (b) Pune-Mumbai (including surrounding areas in Maharashtra); (c) NCR; and (d) Other (other firms located outside the three clusters).

4.1. Variables representing performance

Some studies argue that firms located in a cluster should perform better in terms of profits, growth, and productivity (Helsley and Strange 2001). If agglomeration provides advantages to the sample Indian firms, then these should appear in their performances. In this section, we use two variables as performance indicators: profit margins and labor productivity. All the variables are at the firm level. The sample consists of component manufacturing firms, but excludes car manufacturers, as there are only about a dozen car manufacturers in these clusters.

Profit Margins refers to gross profits as a ratio of sales turnover [(total revenue – total cost)/total revenue]. We expect profit margins to be higher among firms that are located in a cluster.

Labor Productivity refers to the productivity of a rupee spent on labor. Indian firms are not required to disclose the number of employees and consequently their balance sheets do not report employment statistics. Under these circumstances, we suggest that a variable like value added per unit cost of labor, that is, value added divided by the sum spent on labor would be a more appropriate variable, as used in some other studies (Ray 2004; Caves 1992). We assume that firms will be more interested in the productivity of the sums they spend on labor than in an indirect measure of productivity of a representative person they employ.

4.1.2. Conduct variables

Advertisement refers to the advertisement expenditures as a ratio of sales turnover. Since all the sample firms are component manufacturers, we expect established firms in prominent clusters that have durable long-term relationships with vehicle manufacturers/assemblers to advertise less as they do not need to look for customers. On the other hand, firms located outside the clusters need to advertise more to attract customers.

Exports refer to the export intensity or exports to sales ratio. In theory, firms located in clusters are likely to primarily deal with other cluster firms and are expected to have a durable relationship with the vehicle manufacturing firms. Therefore, they may not look actively for the export market. Hence, we expect firms located outside the clusters to export more.

FDI refers to the share of foreign promoter's investment in the firm's total equity capital. This could be interpreted as a measurement of intra-firm transfer of technology.

Inventory refers to the ratio of inventory to sales. We expect firms in clusters that are dominated by multinational vehicle manufacturing firms, like Ashok Leyland, Ford, Hyundai and Hindustan Motors (with Mitsubishi involvement) in Chennai, and MUL and Honda in the NCR, to have lower inventory holdings as these clusters have developed a culture of low inventory and JIT delivery systems. Other firms that do not belong to a cluster might hold larger inventories as they have to ship their products to customers in different regions.

Sales refers to the sales turnover as a proxy for the size of the firm. We expect the cluster firms to be of a larger size and enjoy higher turnovers.

Material Import refers to the import of components and materials as a ratio of sales. It refers to the transfer of embodied technology. Firms that supply to multinationals might use more imported materials as they are likely to face more stringent quality standards and India need not produce all the high-quality materials.

Tech Import refers to the payments of royalty and technology fees as a ratio of sales. We expect technology payments to be higher among firms in clusters where the vehicle manufacturing is dominated by large multinationals as they expect the component firms to keep pace with new technology.

4.2. Econometric analysis

4.2.1. Data set

For empirical analysis we used the *Capital Line* database, one of the most frequently cited firm-level databases available in India, providing data on about 8,000 firms registered in India, including multinationals registered on the Indian stock exchanges. Our sample consists of all the automobile component manufacturing firms listed in the *Capital Line* database, covering the period 1998 – 2005. However, we deleted firms (or observations for certain years) that were producing a very low volume of output or showed zero value addition. We have not used a balanced panel as that does not take into account the entry and exit of firms. In the Indian case it is important to take into consideration both the entry of new firms and the exit of inefficient units. Therefore, we have considered an open panel.

Table 5.3 presents the group means and standard deviations of the firms in the four groups for the performance and conduct variables.

As Table 5.3 shows, some of the Pune-Mumbai firms appear to be engaged in both trading and manufacturing, as their gross profits in relation to sales are very high. Therefore, we concentrate more on the NCR and Chennai clusters and other non-cluster firms.

The mean values of the profit margins are higher for the cluster groups than for the 'other' non-cluster group. This result is valid even if we ignore the Pune-Mumbai cluster. Whether or not this result is statistically significant in a multivariate analysis will be determined in the logistic regressions as discussed below. The results presented in Table 5.3 show that being part of a cluster positively influences the performance of the firms. Their profit margins are more than double those of the non-cluster firms. Furthermore, productivity per rupee spent on labor is also higher for the firms located in the three clusters compared to the non-cluster firms.

Table 5.3 Group mean and standard deviation of the four groups

Variables	Others	Chennai	Pune-Mumbai	NCR
	Mean/SD	Mean/SD	Mean/SD	Mean/SD
Performance				
Profit Margins	0.029/(0.501)	0.069/(0.551)	0.584/(5.571)	0.071/(0.146)
Labor Prod	2.484/(1.813)	2.742/(4.468)	2.762/(1.968)	2.936/(1.694)
Conduct				
Advertisement	0.0070/(0.015)	0.0023/(0.0049)	0.0064/(0.019)	0.0016/(0.0035)
Exports	0.1120/(0.172)	0.0687/(0.123)	0.037/(0.070)	0.051/(0.087)
FDI	11.06/(22.86)	5.319/(12.331)	12.505/(21.214)	10.140/(18.707)
Inventory	0.196/(0.187)	0.152/(0.244)	0.218/(0.630)	0.139/(0.127)
Sales	97.05/(198.90)	138.10/(152.79)	124.38/(168.99)	134.46/(132.47)
Material Import	0.046/(0.069)	0.077/(0.094)	0.057/(0.173)	0.084/(0.107)
Tech Imports	0.003/(0.0068)	0.0049/(0.009)	0.016/(0.150)	0.0049/(0.0079)

Source: author.

With regard to the conduct variables, the table shows that firms in Chennai and the NCR spend much lesser on advertisement (in relation to their sales turnover) compared to the firms in the other two groups. This result is also expected as the firms located in Chennai and the NCR are well-established and well-known and enjoy a multinational clientele. For them additional return from extra advertisements is likely to be low. On the other hand, the firms in the other two groups are relatively new and need to advertise more to establish themselves in the industry.

The non-cluster firms enjoyed higher export intensities compared to the firms located in the three clusters. This result is also expected. The firms in the three clusters supply mainly to car manufacturing units within the cluster and also to other firms located in other clusters. The non-cluster firms need to opt for the export market in order to expand their activities.

Chennai firms have lower levels of FDI (foreign equity participation) compared to those in the other three clusters. We attribute this to historical factors. The main Chennai-based business houses were set up in the late 1950s and early 1960s. By the time India liberalized FDI, the Chennai firms were firmly established and well recognized and could attract foreign strategic non-equity alliances. Thus they had little need to seek foreign equity participation. MNE theories suggest that the transaction costs in transferring technology are higher for investors dealing with relatively new firms compared to established and reputed firms. The theory further suggests that where transaction costs are high, MNEs will prefer FDI to the licensing of technology (Dunning 1993, Teece 1977, Siddharthan and Safarian 1997).

The firms in Chennai and the NCR had lower ratios of inventory to sales compared to the other two groups. This is also anticipated. The need for low inventories and JIT delivery is better appreciated in clusters dominated by vehicle manufacturing multinationals. The main vehicle manufacturers in the Pune cluster are Indian firms. Lower inventories in Chennai and the NCR could be attributed to spillover effects of MNEs.

The non-cluster firms are much smaller than those in the clusters. The average size of the Chennai firm is much larger than that of firms in the other three groups; NCR comes second followed by Pune, with the non-cluster firms coming last. As we described in Section 3, the Chennai firms catered to the all-India market, and they supplied components to all the major vehicle manufacturers located in various parts of the country, including those in the NCR and the Pune-Mumbai clusters. The result confirms our description in Section 3.

The Chennai and NCR clusters import more materials and make more technology payments compared to the other two groups. It is possible that the firms located in these two clusters produce more sophisticated and technology-intensive goods and therefore need to import high-quality materials and frequently upgrade their technology through technical fee and royalty payments. It is also possible that the leading multinationals to which they supply their components – Ford, Hyundai, Hindustan Motors (Mitsubishi) and MUL – require higher standards in terms of technology and sophistication. Low technology payments by component firms in Pune could also be explained by the dominance of the domestic auto manufacturers, such as Tata Motors, which engage in extensive in-house technological development of components, even involving their suppliers. By and large, Table 5.3 indicates that with regard to performance and conduct variables, the Chennai and NCR clusters display similar trends and that these two clusters are very different from the "other" group of firms that do not belong to any cluster. The performance and conduct of the firms in the Pune-Mumbai cluster fall somewhere between those of the non-cluster firms and those in Chennai and the NCR clusters. The firms in the Pune-Mumbai cluster mainly supply components to Indian vehicle manufacturers and that could explain their differential behavior.

4.3. Logistic model and multivariate analysis

In Tables 5.4 and 5.5 in all the equations the dependent variable takes the value of either zero or one. In such cases, Ordinary Least Square estimates are not appropriate and therefore logit or probit models are suggested. In this analysis, we used logit models.[25] The maximum likelihood estimates of the logit model are presented in Tables 5.4 and 5.5.

Table 5.4 presents the results on the inter cluster differences and Table 5.5 on differences between firms that are located inside and outside of clusters. In Table 5.4, Equation 1, firms located in Chennai take the value 1 and those located in the NCR take the value zero. In Equation 2, we assigned the value 1 to firms in Chennai and the value zero to those in Pune-Mumbai. In

Table 5.4 Logit model results: maximum likelihood estimates of inter-cluster differences

	Chennai-NCR	Chennai-Pune	NCR-Pune
Equation	1	2	3
Constant	–0.581	–0.886*	–0.235
Z Stat	(–1.25)	(1.72)	(–0.41)
Profit margin	–.430	–0.450**	–1.579*
	(0.999)	(–2.30)	(–1.72)
Labor Productivity	–0.035	–0.050	0.078
	(–0.87)	(–0.92)	(0.99)
Advertisement	22.841	–68.22***	–63.25***
	(1.05)	(–4.01)	(–3.47)
Exports	1.081	4.943***	4.347***
	(1.25)	(3.27)	(2.89)
FDI	–0.020***	–0.033***	–0.017***
	(3.19)	(–4.57)	(–3.03)
Inventory	1.006	–0.483	–1.040
	(1.26)	(–0.92)	(–0.84)
L Sales	0.099	0.325***	0.116
	(1.08)	(3.06)	(1.08)
Material import	–0.488	8.787***	9.278***
	(–0.52)	(4.27)	(4.70)
Tech imports	11.700	–10.697	–5.631
	(1.03)	(–0.79)	(-0.66)
LR stat	20.09***	94.72***	83.85***
NOBS	541	431	458

Source: author.

Equation 3, we assigned the value 1 to firms inside the NCR and the value zero to those in the Pune-Mumbai cluster.

Similarly, in Equation 1 of Table 5.5, firms in the three clusters (Cls) are assigned the value of 1 and the firms outside the three clusters are assigned the value of zero. Equations 2 to 4 compare firms from each of the three clusters to firms located outside the three clusters. In all the equations, firms outside the three clusters are assigned the value zero and the firms in the respective clusters are given the value one.

As seen from LR statistics, all the equations in both the tables have good fits and they are all significant at the 1 percent level. As anticipated, the results reported in Table 5.4 indicate that firms located in Chennai and the NCR behave similarly. Except for FDI, no other variable was revealed to be statistically significant in the logit analysis. As explained earlier FDI is significant in the equation for historical reasons. Except for this single difference,

Table 5.5 Logit model results: maximum likelihood estimates, cluster and non-cluster differences

	Cls-other	Chennai-other	NCR-other	Pune-other
Equation	1	2	3	4
Constant/Z Stat	−0.104/(−0.28)	−2.025/(−3.89)[a]	−0.512/(−0.96)	−0.464/−0.84
Profit margin	−0.037/(−0.35)	−0.680/(−1.49)	−0.626/(−0.91)	0.754/(1.96)[b]
Labor prod	−0.010/(−0.34)	−0.021/(−0.44)	0.034/(0.44)	−0.051/(−0.71)
Advertisement	−30.35/(−4.22)[a]	−93.09/(−4.78)[a]	−93.15/(−4.37)[a]	−4.621/(−0.68)
Exports	−3.604/(−5.88)[a]	−2.399/(−3.04)[a]	−3.352/(−3.90)[a]	−6.194/(−4.54)[a]
FDI	−0.012/(2.91)[a]	−0.037/(−4.72)[a]	−0.015/(−2.47)[a]	0.004/(0.78)[a]
Inventory	−0.362/(−0.84)	−0.778/(−0.86)	−2.385/(−2.42)[a]	−1.846/(−1.67)[c]
L Sales	0.370/(4.69)[a]	0.636/(5.66)[a]	0.297/(2.81)[a]	0.191/(1.89)[b]
Material import	4.870/(3.75)[a]	7.923/(4.41)[a]	7.426/(4.54)[a]	1.329/(0.60)
Tech imports	20.885/(1.81)[c]	31.867/(1.65)[c]	27.666/(1.70)[c]	26.622/(1.88)[b]
LR stat	118.45[a]	128.64[a]	134.44[a]	58.25[a]
NOBS	946	489	516	406

Note: a, significant at 1%, b, at 5% and c, at 10% levels.

Source: author.

the two clusters appear to be homogeneous. The results for equations 1, 2 and 3 in Table 5.4 are similar, further reinforcing the similarity between the Chennai and NCR clusters, and the difference compared to those firms located outside the clusters. Equations 2 and 3 in Table 5.4 clearly show that firms in the Pune-Mumbai cluster behave differently from those in the Chennai and NCR clusters.

Table 5.5 shows that the performance variables—profit margins and productivity—do not significantly differentiate the firms belonging to the three clusters from those located outside the clusters, in the presence of the conduct variables. It is quite possible that firms located in these two clusters (Chennai and the NCR) enjoyed higher scores on the performance indicators mainly because of their conduct variables but when they are introduced directly, the performance variables turn out to be insignificant. Thus, we can interpret that the Chennai and NCR firms enjoyed better profit margins and productivities mainly because they used better material and constantly upgraded their technology. However, when these two variables were used in the equation along with the performance variables, the performance variables lost their significance.

Equations 2 and 3 in Table 5.5 show that firms located in Chennai and the NCR clusters spend much less on advertisements compared to the firms that are not part of the three clusters. Furthermore, firms in Chennai and the NCR spend much less on advertisements than the firms located in Pune-Mumbai. In addition, the firms in these two clusters export less compared to the firms located in Pune-Mumbai and the non-cluster firms. These two variables—advertisement intensity and export intensity—are significant in all the relevant equations at the 1 percent level. Likewise, FDI is also statistically significant in all the equations in Table 5.5 except Equation 4. Inventory sales ratio is significant only in Equation 3, indicating that firms in the NCR may hold fewer inventories compared to those outside the clusters, suggesting the diffusion of the JIT principle across the cluster firms. In Table 5.3, Chennai firms also held smaller inventories but it has not turned out to be important in the multivariate analysis. The average size of the firm as represented by log scales has also been shown to be significant at the 1 percent level in all the three equations (2, 3 and 4) that differentiate the non-cluster firms from the three respective clusters. Thus, the larger size of the firms located in the three clusters, which we observed from Table 5.3, has continued to be significant in Table 5.5. Table 5.5 further shows that the firms in Chennai and the NCR are more import intensive (imports of components and materials) than those in the other two clusters. This variable is also significant at the 1 percent level. However, when it comes to technology imports, it is significant at only the 10 percent level. In other words, results presented in Table 5.5, by and large, confirm all the findings of Table 5.3, except the performance variables.

In sum, clustered firms are larger in size, maintain smaller inventories, import more materials and components, and spend more on technology imports. However, they are less export intensive and spend less on advertisement. Since the cluster firms are large and enjoy a durable relationship with the vehicle assemblers, they need not spend much on advertisement and need not depend on exports. Furthermore, since the main customers are MNEs (also located in the cluster), the cluster firms need to import better-quality materials and update their technology frequently, which means incurring technology payments. Thus, the overall results support our hypothesis and show that the behavior of firms in clusters differs significantly from that in non-cluster firms.

5. Conclusion and main lessons of the study

Our study of the automobile clusters in Chennai and the NCR reveals interesting differences in the patterns of cluster formation and growth. We find that firms in the Indian auto industry are largely located in three main clusters: Chennai (Tamil Nadu), Pune-Mumbai (Maharashtra), and the national capital region (NCR), as in some other modern manufacturing sectors. All these clusters are thus multi-industry clusters and their continued growth supports Jacobs's (1969) emphasis on urban diversity in contrast to urban specialization.

The formation of the Chennai cluster in the late 1950s and the 1960s was driven mainly by large component manufacturers, many of whom were member firms of large Indian business houses. Agglomeration of auto component manufacturers occurred largely because Chennai had access to seaports, to a pool of an educated workforce, as well as the strong leadership in the state government who actively promoted the industry in the region. Access to seaports was critical in its initial years, due to the industry's heavy reliance on imported materials and parts.

While conforming to his capacity building argument, Chennai's experience differs considerably from Kuchiki's (2004) model with respect to the role of the anchor firms. The component manufacturing firms that came first in the late 1940s did not heavily depend on the two vehicle manufacturers that existed in the cluster in their initial years. From inception, the component manufacturers have targeted the all-India market, due to a small volume of production in the industry as a whole. Thus, due to their small production volume, these two vehicle assemblers played very limited roles in developing the cluster. The key component manufacturers also avoided joint ventures and FDI participation in the subsequent years because of the Indian government's policy that restricted FDI until the mid-1980s. Thus, key anchor firms in the Chennai auto cluster have been mainly component manufacturers that were established in the late 1950s and the 1960s, rather than assemblers.

Therefore, because its formation was led by local component manufacturers, including many small firms, the Chennai cluster is quite unique, compared to other auto clusters in India or elsewhere. This pattern also differs considerably from the "flowchart model" presented by Kuchiki (2004). In his model, industrial parks and zones first attract lead firms to locate there, and then they play a central role in bringing in related firms as the cluster develops its capacity in terms of human, physical and institutional infrastructures. The opposite occurred in the Chennai cluster: assembler firms began to play leading roles in cluster development only in the late 1990s, when global players like Hyundai, Ford, and Mitsubishi came to Chennai to start their production. The presence of many component manufacturers with competent engineering skills in this cluster attracted these global car makers to locate there.

By contrast, the auto cluster in the NCR was created mainly by a single assembler firm, MUL, as a lead firm. Thus, the pattern of this cluster generally confirms Kuchiki's (2004) model. The related firms, i.e., auto component suppliers and dealers, were actively developed by the anchor firm. Various policy factors have motivated MUL to actively develop its first-tier suppliers, such as the requirement to increase local content, high duties imposed on imported parts, and the reservation of various items for production by domestic small-scale firms. Moreover, as it is the national capital, the NCR enjoys the closest links with the centers of power and patronage; this has favored MUL, with which the central government is a joint venture partner. Thus, the contrasts between the Chennai and NCR clusters in the patterns of their formation suggest the presence of inter-cluster variations even within the same industry, and that such inter-cluster variations are partly explained by the historical and policy conditions under which firms, particularly lead firms, must operate.

On the other hand, the two clusters share some common features, such as the creation of industrial zones and the availability of high levels of skilled workers in these clusters. In both clusters, the state government actively supported the creation of industrial estates and the development of infrastructure. Interestingly, unlike the experiences in Viet Nam and China (Kuchiki 2004), in these Indian auto clusters, the state governments developed such industrial estates, not to attract FDI but to primarily promote domestic firms, and in particular, small firms.

The study also brings to light other aspects of the state's role: industrial licensing and location policies, and the role of political leaders in influencing industrial location. Until 1991, industrial location was not a free choice of firms guided by commercial consideration. The industrial licensing policy decided on the location of a particular industry at the state level, and the state government in turn influenced the decisions about locations within the state.

Furthermore, the study reveals the inter-dependence of, and inter-relationships between, industries located in a cluster. In Chennai, IT firms located in the cluster have helped the small and medium-sized automotive firms become globally competitive. Thus, multi-industry clusters have an advantage.

Our econometric analyses confirmed that being part of a cluster positively influences the performance of the auto component firms and that firms located in clusters behave differently from and perform better than those firms located outside the clusters, as indicated by the conduct and performance variables. These analyses clearly suggest the advantage to firms in a cluster over those outside clusters.

Acknowledgments

We thank the IDE for its financial support to carry out this research. We also thank Mr. Gubbala Lakshmana Rao for his research assistance.

Notes

1. The National Capital Region (NCR) is the area in and surrounding the national capital Delhi. It includes some districts in the state of Haryana (such as Gurgaon and Faridabad), and some in Uttar Pradesh (such as NOIDA and Ghaziabad) that are adjacent to Delhi.
2. This is one of the most widely-used databases in India, available on line as well as on CD by subscription. It provides data on about 8000 firms registered in India, including multinationals registered in the Indian Stock Exchanges.
3. See also Kuchiki (in this volume).

4. See Markusen (1996) for the typology of industrial agglomerations. See also Fujita (in this volume) for discussions on Markusen's typology.
5. Initially, TELCO entered into a joint venture with Mercedez-Benz, as did half of the 120 Tata-affiliated firms, in the late 1950s, but the partnership expired in 1962 (Encarnation 1989).
6. The government restricted the inflow of foreign capital, and the Foreign Exchange Regulation Act (FERA) of 1974 forced foreign investors to keep their share to 40% or less (Lall 1987). For detailed discussions of this policy and its effects, see Lall 1987.
7. T.T. Krishnamachari was from Tamil Nadu and occupied some key cabinet ministerial positions, including those of Finance, Commerce and Industry, in the Government of India, and played a key role in the development of auto industry in Tamil Nadu.
8. They include T. V. Sundaram Iyengar (TVS group); Mammen Mapillai (MRF); Raghubir Saran (Ashok Leyland); Gopalakrishnan of Union Motors (Standard 10, Standard Companion, and Standard Herald and Gazel); N. Mahalingam, (Anamalai Body Building Works for trucks and buses and Anamalai Tyre Retreading); L. G. Brothers (body building and auto equipment); L. L. Narayanan (Rane Group, producing steering wheels and brake linings); and Anantharama Krishnan (India Pistons, Tractors and Farm Equipment and SRVS).
9. Even today tyres for Rolls-Royce cars are produced at MRF.
10. TVS & Sons Ltd., the parent and holding company of the TVS group, was established in 1911 and has now become the largest automobile distribution company in India with a turnover of more than US $450 million.
11. The firm was established in 1982 in technical and financial collaboration with Aktiengesellschaft Kuhnle, Kopp & Kausch, Germany. Its manufacturing plant is located in Chennai and Vellore (Tamil Nadu), and its R&D unit in Padi, Chennai.
12. The Hinduja Group and IVECO, Italy (a subsidiary of Fiat) acquired Leyland, UK in 1987: this made Land Rover Leyland International, UK, the holding company of ALL. The holding company holds 50.9% of stake in ALL's equity.
13. Hindustan Motors' Chennai plant is engaged in manufacturing of Lancer cars in equity and technical partnerships with Mitsubishi Motors, and in manufacturing and sales of the spare parts used in the Lancer cars. In response to the entry of many global players to the Indian market, domestic car producers such as Hindustan Motors and Premier Auto Ltd. (PAL) formed financial partnerships with General Motors and Mitsubishi, and Fiat and Peugeot respectively. Hindustan Motors earlier formed technical collaboration with Isuzu Motors.
14. After the new economic and industrial policies of 1991, in its interest to promote FDI the Indian government deliberately developed no specific policy on the automobile sector with respect to local content. While its Phased Manufacturing Program (PMP) was lifted in 1992, however, the government still implicitly demanded 50 percent local content in approving foreign collaboration proposals in the 1990s; this would rise to 70 percent after five years, often specified in "Memorandum of Understanding" (MOU) signed with each new entrant on a case-by-case basis (Humphrey et al. 1998; Okada 2000).
15. Initially, Sanjay Gandhi, the elder son of then-Prime Minister Indira Gandhi, started Maruti Ltd., as a private firm to achieve his dream of producing a "national car." Due to some political problem, however, it never started operation. Indira Gandhi's government nationalized the firm in 1980, after her son's death, to

achieve his dream. Advised that the project would not succeed without the involvement of foreign technology, the government decided to engage in a joint venture, and signed joint-venture and license agreements with Suzuki in 1982 (interview with a former CEO, Maruti Udyog Ltd., at Harvard University, February 1998). MUL remained as a state enterprise until 1992, when the government reduced its share of equity from 60% to 49.9%, in accordance with government policy change that allowed state enterprises to form joint ventures (Okada 2000). However, the government recently announced its policy to disinvest from MUL, which will allow it to become Suzuki's subsidiary firm.

16. MUL's Annual Report 1996–97.
17. This system differs from the Toyota style, where most component manufacturing firms either belonged to Toyota *keiretsu* or received technology from the Toyota related firms (Hattori 1996). This is because Suzuki, as a relatively small assembler in Japan, has a relatively small supplier base at home, compared to other larger assemblers such as Toyota and Nissan. It has a *"kyoryokukai"* (supplier association formed by a car manufacturer) only for the small suppliers located near its plants (Fujimoto and Takeishi 1994). When MUL started its operation in India, few dedicated suppliers of Suzuki could afford to follow Suzuki to India to supply parts for MUL (Okada 2000).
18. Data are from MUL's supplier database (1997). Thus, there might be some false reporting on the part of suppliers to indicate their loyalty to MUL. Therefore, the actual dependency could be lower.
19. Data are from MUL's supplier database (1997).
20. The price of CKD was determined in US dollars, and thus the exchange rate between the US dollar and the Japanese yen significantly affected the cost of production (talk by Mr. R.C. Bhargava, former CEO of MUL, at Harvard University, February 1998).
21. See Fujimoto and Takeishi (1994) for the figure on Japan.
22. In fact, Japanese managers think that the Indian automobile industry, including MUL, is still far from operating under the JIT principle (interviews with senior managers of MUL, Denso India, and the CEO of Toyota India Corp.).
23. As of 1997, more than 64 auto components were reserved to be produced by the SSI sector. Even after liberalization in 1991, the reservation policy for SSI continued, with only 7 items removed from the list by 1994 (Okada 2000).
24. See Okada (2000) for more detailed discussions on MUL's training for its employees and its suppliers.
25. Probit models also yielded similar results.

Bibliography

Acs, Z. J., D. B. Audretsch and M. P. Feldman. 1994. "Spillovers and Recipient Firm Size." *Review of Economics and Statistics*, Vol. 76, no. 2: 336–40.

Asheim, B. T. and L. Coenen. 2005. "Knowledge Base and Regional Innovation Systems: Comparing Nordic Clusters." *Research Policy*. Vol. 34 No. 8: 1173–1190.

Association for Indian Automobile Manufacturers (AIAM). 1997. *The Automobile Industry: Statistical Profile 1997*. Bombay: AIAM.

Audretsch, D. B. and M. P. Feldman. 1996. "R&D Spillovers and Geography of Innovation and Production." *American Economic Review*. Vol. 86 No. 3: 630–640.

Audretsch, D. B. and E. E. Lehmann. 2005. "Does the Knowledge Spillover Theory of Entrepreneurship Hold for Regions?" *Research Policy*. Vol. 34 No. 8: 1191–1202.

Baldwin, R. E. and P. Martin. 2004. "Agglomeration and Regional Growth." in J. V. Henderson and J. F. Thisse. *The Handbook of Regional and Urban Economics*. Vol. 4. Amsterdam: Elsevier, 2672–2711.

Belderbos, R. and M. Carree. 2002. "The Location of Japanese Investments in China: Agglomeration Effects, Keiretsu, and Firm Heterogeneity." *Journal of the Japanese and International Economies*. Vol. 16 No. 2: 194–211.

Braunerhjelm, P. and R. Swenson. 1996. "Host Country Characteristics and Agglomeration in Foreign Direct Investment." *Applied Economics* Vol. 28: 833–40.

Breschi, S. and F. Malerba. 2001. "The Geography of Innovation and Economic Clustering: Some Introductory Notes." *Industrial and Corporate Change*. Vol. 10 No. 4: 817–33.

Breschi, S. and F. Lissoni. 2001. "Knowledge Spillovers and Local Innovation Systems: A Critical Survey." *Industrial and Corporate Change*. Vol. 10 No. 4: 975–1005.

Caves, R. E. 1992. *Industrial Efficiency in Six Nations*. Cambridge, Massachusetts: MIT Press.

Caves, R. E. 1996. *Multinational Enterprise and Economic Analysis*. (Second Edition), Cambridge Survey of Economic Literature, Cambridge: Cambridge University Press.

Cooke, P. 2001. "Regional Innovation Systems, Clusters, and the Knowledge Economy." *Industrial and Corporate Change*. Vol. 10 No.4: 945–974.

Dunning, J. H. 1993. *Multinational Enterprises and the Global Economy*. Reading, UK: Addison-Wesley Publishing Company.

Eaton, C. B., R. G. Lipsey and A. E. Safarian. 1994. *The Theory of Multinational Plant Location: Agglomeration and Disagglomerations*. Economic Growth and Policy Program, Canada Institute of Advanced Research, Canada.

Encarnation, D. J. 1989. *Dislodging Multinationals: India's Strategy in Comparative Perspective*. Ithaca, NY: Cornell University Press.

Feldman, M. P. 1994. "Knowledge Complementarity, and Innovation." *Small Business Economics*. Vol. 6 No. 5: 363–72.

Ford, S. and R. Strange. 1999. "Where do Japanese manufacturing firms invest within Europe, and why?," *Transnational Corporations*, Vol. 8, Part 1, pages 117–142.

Fujimoto, T. and A. Takeishi. 1994. *Jidosha Sangyo 21 Seiki heno Scenario: Seishou gata System kara Balance gata System heno Tenkan (Towards the "lean-on-balance" System)*. Tokyo: Seisansei Shuppan.

Fujita, M., P. Krugman and A. J. Venables. 1999. *The Spatial Economy: Cities, Regions, and International Trade*. Cambridge, MA: MIT Press.

Globerman, S. 2002. "Global Foreign Direct Investment Flows: The Role of Governance Infrastructure." *World Development*. Vol. 30, No.11: 1899–1919.

Grossman, G. M and E. Helpman. 1991. *Innovation and Growth in the Global Economy*. Cambridge, MA: The MIT Press.

Gulyani, S. 2001. "Effects of Poor Transportation on Lean Production and Industrial Clustering: Evidence from the Indian Auto Industry." *World Development*. Vol. 29 No. 7: 1157–77.

Hattori, T. 1996. "Technology Transfer and Industrial Cooperation between India and Japan: Maruti and Suzuki." A paper presented at a conference at Hitotsubashi University, Tokyo, March 8–9.

He, C. 2002. "Information Costs, Agglomeration Economies and the Location of Foreign Direct Investment in China." *Regional Studies*. December Vol. 36 No. 9: 1029–36.

Head, K., J. Ries and D. Swenson. 1995. "Agglomeration Benefits and Location Choice: Evidence from Japanese Manufacturing Investments in the United States." *Journal of International Economics*. Vol. 38: 223–247.

Head, K. and J. Ries. 1996. "Inter–city Competition for Foreign Investment: Static and Dynamic Effects of China's Incentive Areas." *Journal of Urban Economics*. Vol. 40 No. 1: 38–60

Helsley, R. W. and W. C. Strange. 2001. "Innovation and Input Sharing." *Journal of Urban Economics*. Vol. 51 No. 1: 22–45.

Henderson, J. V. 1986. "Efficiency of Resource Usage and City Size." *Journal of Urban Economics*. Vol. 19: 47–70.

Henderson, J. V. 2003. "Marshall's Scale Economies." *Journal of Urban Economics*. Vol. 53: 1–28.

Humphrey, J., Mukherjee, A., Zilbovicius, M. and Arbix, G. 1998. "Globalization, FDI and the Restructuring of Supplier Networks: the Motor Industry in Brazil and India," in M. Kagami, Humphrey, J. and Piore, M. (eds), *Learning, Liberalization and Economic Adjustment*. Chiba: Institute of Developing Economies, 117–189.

Jacobs, J. 1969. *The Economics of Cities*. New York: Vintage.

Jianping, Ding. 1999. "Agglomeration Effects in Manufacturing Location – Are There Any Country's Preferences?' *Economia Internazionale*. Vol. 52 No. 1: 59–78.

Karlsson, C., B. Johansson and R. R. Stough. (eds) 2005. *Industrial Clusters and Inter-firm Networks*. Cheltenham: Edward Elgar Publishing.

Kathuria, V. 2000. "Productivity Spillovers from Technology Transfer to Indian Manufacturing Firms." *Journal of International Development*. Vol. 12: 343–369.

Kathuria, V. 2002. "Liberalisation, FDI and Productivity Spillovers – an Analysis of Indian Manufacturing Firms." *Oxford Economic Papers*. Vol. 54: 688–718.

Krugman, P. 1991. "Increasing Returns and Economic Geography." *Journal of Political Economy*. Vol. 99 No. 3: 483–499.

Kuchiki, A. 2004. "A Flowchart Approach to Asia's Industrial Cluster Policy," in Akifumi Kuchiki and Masatsugu Tsuji (eds). *Industrial Clusters in Asia: Analyses of Their Competition and Cooperation*. IDE Development Perspective Series No.6, Chiba, Japan: Institute of Developing Economies, 193–225.

Lall, S. 1987. *Learning to Industrialize: The Acquisition of Technological Capability in India*. Basingstoke and London: Macmillan.

Lall, S. and Siddharthan, N. S. 1982. "The Monopolistic Advantages of Multinationals: Lessons from Foreign Investment in the US." *Economic Journal*. Vol. 92: 668–683.

Markusen, A. 1996. "Sticky Places in Slippery Space: A Typology of Industrial Districts." *Economic Geography*. Vol. 72: 293–313.

Marshall, A. 1920. *Principles of Economics*. London: Macmillan.

Nakamura, R. 1985. "Agglomeration Economies in Urban Manufacturing Industries: A Case of Japanese Cities." *Journal of Urban Economics*. Vol. 17: 108–124.

Okada, A. 2000. *Workers' Learning through Inter-firm Linkages in the Process of Globalization: Lessons from the Indian Automobile Industry*. Ph.D. Dissertation. Cambridge, MA: Massachusetts Institute of Technology.

Okada, A. 2004. "Skills Development and Interfirm Learning Linkages under Globalization: Lessons from the Indian Automobile Industry." *World Development*. Vol. 32 No.7: 1265–1288.

Ramachandran, V. and J. Goebel. 2002. *Foreign Direct Investment in Tamil Nadu: Review and Comparison across Host Sites*. Harvard University, Website, 1–71.

Ray, S. 2004. "MNEs Strategic Alliances and Efficiency of Firms: Emerging Trends in Liberalizing Era." *Economic and Political Weekly*. Vol. 39 No. 5: 434–440.

Romer, P. M. 1986. "Increasing Returns and Long-run Growth." *Journal of Political Economy*. Vol. 94 No. 5: 1002–1037.

Ronde, P. and C. Hussler. 2005. "Innovation in Regions: What Does Really Matter?" *Research Policy*. Vol. 34 No. 8: 1150–1172.

Rosenthal, S. S. and W. C. Strange. 2003. "Geography, Industrial Organization and Agglomeration." *Review of Economics and Statistics*. Vol. 85 No. 2: 377–393.

Rosenthal, S. S. and W. C. Strange. 2004. "Evidence on the Nature and Sources of Agglomeration Economies." in Henderson, J. V. and J. F. Thisse (eds) *Handbook of Regional and Urban Economics*. Volume 4, Amsterdam: Elsevier B. V. 2120–2171.

Saxenian, A. 1994. *Regional Advantage: Culture and Competition in Silicon Valley and Route 128*. Cambridge, MA: Harvard University Press.

Siddharthan, N. S. and A. E. Safarian. 1997. "TNCs, Technology Transfer and Import of Capital Goods: Recent Indian Experience." *Transnational Corporations*. Vol. 6 No. 1: 31– 49.

Society of Indian Automobile Manufacturers (SIAM). 2002. *The Indian Automobile Industry: Statistical Profile 2000–2001*. Delhi: SIAM.

Sridhar, V. 2002. "Happening Chennai." *Frontline*. Vol. 19 No. 21. October 12.

Syamwil, I. B. and P. H. Tanimura. 2000. "The Spatial Distribution of Japanese Manufacturing Industries in Indonesia." *Review of Urban and Regional Development-Studies*. July. Vol. 12 No. 2: 120–136.

Teece, D. J. 1977. "Technology Transfer by Multinational Firms." *Economic Journal*. Vol. 87: 242–261.

Tewari, M. 2003. *Foreign Direct Investment and the Transformation of Tamil Nadu's Automotive Supply Base*. University of North Carolina at Chapel Hill, Website, http://www.ids.ac.uk/globalvaluechains/publications/AutoLoraine2003-FINAL.pdf. 1–41.

Tuan, C. Ng. and F. Y. Linda. 2003. "FDI Facilitated by Agglomeration Economies: Evidence from Manufacturing and Services Joint Ventures in China." *Journal of Asian Economics*. January. Vol. 13 No. 6: 749–65.

Wei, Y. 1999. "The Regional Distribution of Foreign Direct Investment in China." *Regional Studies*. Vol. 33 No. 9: 857–67.

Zucker, L. G., M. R. Derby and M. B. Brewer. 1998. "Intellectual Property and the Birth of US Biotechnology Enterprises." *American Economic Review*. Vol. 88 No. 1: 290–306.

6

The Process and Factors of Industrial Cluster Formation: A Flowchart Approach to Industrial Cluster Policy in Japan

Kentaro Yoshida

1. Introduction

Throughout the early 1990s, the chief aim of Japan's industrial policy with regards to outlying economies (i.e. areas outside the major cities) was to promote local development in those areas, by relocating various industries out of the major cities and by creating industrial agglomerations. However, in the late 1990s, so-called "deindustrialization" accelerated, as factories began leaving outlying areas and relocating overseas in reaction to the stronger yen and the industrialization of China and ASEAN. This made it impossible for Japan's outlying economies to attract companies from large cities.

In response to this situation, the idea of industrial clusters, a new approach to economic growth, has attracted recent attention. The basic idea is to replicate the successes of the mid-1990s, the best-known of which are Silicon Valley and Boston, by using policy to concentrate strategic industries in specific regions, in order to develop not only those local economies but also the national economy overall (Saxenian 1994). The goal is to boost the competitiveness of both companies and regions, by supplementing growth-promoting macro-economic policy with a micro-economic approach. The fundamental feature of industrial clusters is that new added value is created through innovations that arise from collaboration, competition, and cooperation among a diverse array of organizations and institutions in a region where not only companies and industries but also government and academia are represented (M. Porter 1998).

We thus seek a theoretical basis for determining the prerequisites for cluster formation.

Porter (1998) analyzed the development factors that create industrial clusters using a "diamond approach," but did not present industrial clusters as a component of policy. Kuchiki (2003) applied a "flowchart approach" to clusters in northern Vietnam and demonstrated the effectiveness of official development assistance (ODA). Kuchiki (2004) applied the flowchart approach to Tianjin, China, analyzing the factors obstructing industrial agglomeration there. Kuchiki (2005A) presented the flowchart approach to industrial cluster policy in its entirety, and demonstrated the effectiveness of capacity building and invitation of anchor firms as part of cluster policy. However, there as yet exists no quantitative analysis of the factors affecting cluster development. Moreover, although flowcharts of industrial agglomerations have been presented, there are as yet no flowcharts that deal with the promotion of the innovations that are indispensable for agglomerations to develop into true industrial clusters.

On the other hand, the research[1] of Ishikura et al. (2003) on industrial clusters in Japan, which illustrated the significance of planned cluster development in Japan, positioned cluster theory within the broad scope of the new field of spatial economics, as well as within strategic theory and organization theory from business administration studies. Yamazaki (2002), who clarified the characteristics and the policy issues of Japan's clusters by comparing Japan's advanced clusters with those in the US, Finland, and China, emphasized the necessity of shifting to a new regional strategy that utilizes the industrial agglomerations and know-how agglomerations that have been nurtured in various regions in Japan. Itami, Matsushima, and Kikkawa (1998) demonstrated the significance of Japan's industrial agglomerations as sites that accumulate resources for the market, taking into account theories of industrial agglomerations as well as an analysis of the current state of industrial agglomerations. However, these studies make no reference to practical industrial cluster policies.

The objective of this chapter is to analyze Japan's industrial cluster policy, evaluate the clusters that have formed in Japan as the result of that policy, define the process by which these clusters formed, and identify the factors that promoted that process.

We first present the current status of various types of industrial clusters across Japan, most of which were formed as a result of policy decisions. We then compare and contrast the development of the various cluster types. We present a hypothesis regarding how and why clusters form, and we test this hypothesis with a questionnaire survey of Japanese cluster policymakers. Finally, we present these factors in a flowchart model and suggest a practical policy for industrial clusters.

Section 2 discusses the history of industrial policy and industrial cluster policy in Japan, and the theory that provides the context for cluster policy. Section 3 introduces industrial cluster projects that are currently being developed in Japan, and presents our hypothesis of cluster formation.

Section 4 presents and analyzes the results of a questionnaire survey based on the above-mentioned hypothesis. Section 5 relates the conclusions gleaned from that analysis and suggests directions of further study. An outline of the questionnaire survey is found in the appendix.

2. Industrial cluster policy in Japan

2.1. Industrial policy, present and future

Industrial cluster policy was first introduced in Japan in 2001, to boost the competitiveness of Japanese industries in the international market and to help local economies become independent through self-development. Japan's Industrial Cluster Plan has two components: the Industrial Cluster Plan itself, which is promoted by METI, and the Project for Creating Knowledge Clusters, promoted by the Ministry of Education, Culture, Sports, Science, and Technology. The overall objective is to create an optimal environment for innovation, with an emphasis on forming networks in each region between industry, government, and academia. The effect of these networks, it is hoped, will be to energize local economies by facilitating the production of new technologies and industries.[2]

METI is now implementing the following industrial cluster projects throughout Japan:

1. Hokkaido region:
    - Strategy for the Promotion of the Hokkaido Super Cluster
    - Bio and Information Industry Cluster
2. Tohoku region (Aomori, Yamagata, Iwate, Akita, Miyagi, Fukushima):
    - Industrial Project for Creation of Information – Bioscience and the Future
3. Tohoku region (Aomori, Yamagata, Iwate, Akita, Miyagi, Fukushima):
    - Project for Promoting the Industry of a Recycling Society
4. Kanto region (Ibaraki, Tochigi, Gunma, Saitama, Chiba, Tokyo, Kanagawa, Nigata, Yamanashi, Nagano, Shizuoka):
    - Project for Regional Revitalization
5. Capital region (centered on Tokyo and Yokohama and including Kazusa, Tsukuba and Kashiwa provinces):
    - Fostering Bio-Ventures: the Capital Region Bio Genome Network
6. Capital region (Tokyo, Kanagawa, Saitama, and Chiba prefectures):
    - Capital Region Information Venture Forum
7. Tokai region (Aichi, Gifu and Mie prefectures):
    - Project to Create New Manufacturing in Tokai
8. Tokai region (Aichi, Gifu and Mie prefectures):
    - Project to Create Bio Manufacturing in Tokai

9. Hokuriku region (Toyama, Ishikawa and Fukui prefectures):
   - Project to Create New Manufacturing in Hokuriku
10. Kinki region (Fukui, Shiga, Kyoto, Osaka, Hyogo, Nara and Wakayama prefectures):
    - Kinki Bio-Related Industry Project
11. Kinki region (Fukui, Shiga, Kyoto, Osaka, Hyogo, Nara and Wakayama prefectures):
    - Project for Promoting Lively Manufacturing
12. Kinki region (Fukui, Shiga, Kyoto, Osaka, Hyogo, Nara and Wakayama prefectures):
    - Project for Promoting an Information-Related Cluster
13. Kinki region (Fukui, Shiga, Kyoto, Osaka, Hyogo, Nara and Wakayama prefectures):
    - Project for Promoting Advanced Use of Energy and the Environment
14. Chugoku region (Tottori, Shimane, Okayama, Hiroshima, Yamaguchi):
    - Project for Creating a Recycling and Environment-Conscious Society
15. Chugoku region (Tottori, Shimane, Okayama, Hiroshima, Yamaguchi):
    - Project to Create Next-Generation Industry Nuclei
16. Shikoku region (Kagawa, Kochi, Ehime, Tokushima):
    - Shikoku Techno-Bridge Plan
17. Kyushu region (Fukuoka, Saga, Nagasaki, Kumamoto, Miyazaki, Oita, Kagoshima):
    - Kyushu Exchange Plaza for Environment-Related and Recycling-Related Industries
18. Kyushu region (Fukuoka, Saga, Nagasaki, Kumamoto, Miyazaki, Oita, Kagoshima entire region):
    - Kyushu Silicon Cluster Plan
19. Okinawa region:
    - Promotion of Okinawa-Type Industries

METI defines an industrial cluster as "a region in which a business environment favoring innovation is produced by a horizontal network of collaboration among industry, government, and academia as well as collaboration among companies, such that new businesses utilizing each other's enterprise resources are produced in rapid succession, resulting in an industrial agglomeration that progresses with the relatively predominant companies as the core." The core of the policy is the "creation of new industries and new businesses by promoting local innovation, concurrently with the formation of a network of industry, government, and academic collaboration as well as industry-industry collaboration, with the ultimate aim of forming an industrial cluster."[3]

The three aims of Japan's industrial cluster policy are: creation of a business environment where innovations will be produced by so-called "new fusions"; a national strategy of development of, or discovery and

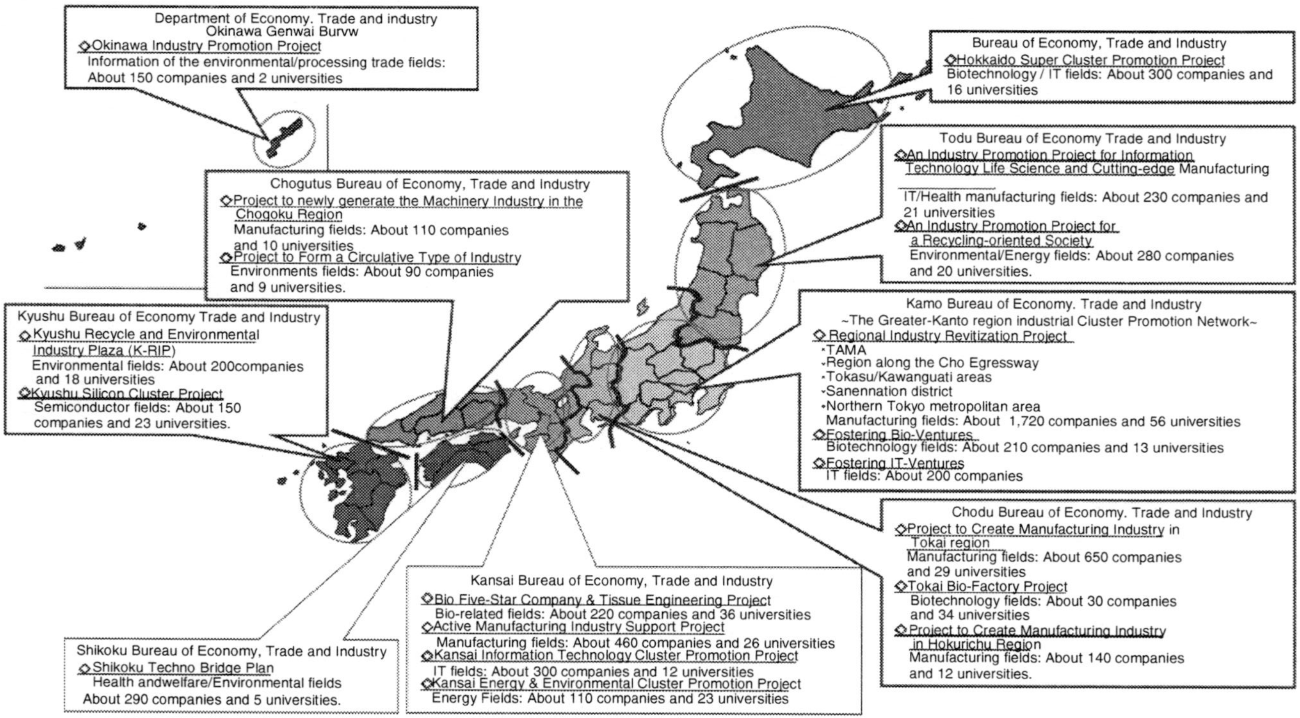

Figure 6.1 Industrial cluster project map in Japan

Source: METI.

backing of, important new industries in outlying areas; and creation of synergistic effects by linking these efforts with local industry promotion implemented by local parties.

2.2. Basic structure of industrial clusters and the roles of policy

Michael Porter, the Harvard University professor who popularized the concept of industrial clusters, defines the concept of cluster thus: "A cluster is a geographically proximate group of interconnected companies and associated institutions in a particular field, linked by commonalities and complementation."[4] In other words, an industrial cluster is an entity in which geographically adjacent agents – companies, universities and research institutions, industrial support agencies, networking organizations, technology transfer agencies, intermediary agencies for industrial-academic collaboration, specialist groups, etc. – draw on a region's unique advantages to coalesce into a mutually interrelated industrial group.[5]

There are three reasons why industrial clustering is a significant phenomenon. The first reason is that clusters generate positive economic externalities by strengthening complementary industries, related institutions, and systems in a single geographic area. The second reason is that a horizontal network of industry, academia, and government has a synergistic effect on the process of innovation, allowing the creation of new businesses. The third reason is that clusters further accelerate the agglomeration of the industry in which they form, drawing together an ever-increasing pool of human resources, firms, and investment.

Policies encouraging the formation of such clusters generally fall into two types. The first involves supporting the growth of local anchor firms and nurturing human resources. The second type of policy involves forming networks among industrial, government, and academic agents to promote R&D projects, or supporting collaborations between diverse industries.

3. Cluster policy and results of industrial cluster formation in Japan

3.1. Government strategy and current condition of Japan's industrial clusters

As part of METI's industrial cluster plan, the ministry's regional bureaus and private promotional organizations have joined together in a nationwide project to construct close cooperative relationships between researchers at over 220 universities and 5,800 small and medium enterprises (SMEs) that are attempting to launch new businesses. The project spent a total of 49 billion yen in 2004, including:

- 4 billion yen for formation of industry, government, and academic networks in outlying areas,
- 38.5 billion yen for promoting development of technologies by utilizing the special characteristics of outlying areas, and
- 6.5 billion yen for strengthening incubation services, including the building of entrepreneur training facilities.

Because this plan has only recently been implemented, its results are not yet well-known; however, because the intermediate results of this program are important for future strategy, we here examine industrial statistics related to the clusters.

Table 6.1 displays statistics regarding the growth in outlying areas that have implemented industrial cluster policies in recent years.[6] Comparison of the statistics for 2002 and 2004 show positive growth in "industrial shipment amount" and "added value." The overall growth rate of "industrial shipment amount" in these areas was 1.06%, and 1.05% for "added value." We see that the outlying areas that display higher growth rates than these are the Tohoku Cluster (manufacturing, recycling), where both "industrial shipment amount" and "added value" growth rates are 1.07%, the Tokai Cluster (manufacturing, biotechnology), where "industrial shipment amount" grew by 1.08% and "added value" by 1.07%, the Chugoku Cluster (manufacturing, recycling) where "industrial shipment amount" grew by 1.10% and "added value" by 1.07%, the Shikoku Cluster (manufacturing) where "industrial shipment amount" grew by 1.08% and "added value" by 1.09%, and the Kyushu Cluster (semiconductors, environmental technology, recycling) where "industrial shipment amount" grew by 1.07% and "added value" by 1.07%.

3.2.1. Manufacturing sector

There are nine cluster projects involving manufacturing, more than any other sector. These projects are located in Tohoku, Kanto, Tokai, Hokuriku, Kinki, Chugoku, Shikoku, Kyushu, and Okinawa. Our discussion will focus on the New Project for the Chugoku Region Machinery Industry in the Chugoku region, where growth has been relatively high in recent years. We examine the characteristics of the manufacturing sector in this area, and the factors affecting cluster formation.

Factors affecting cluster formation

The Chugoku region has a history of development through agglomeration of the machinery and automobile industries. The region's companies boast a wide range of machine metal processing technology, including metal molding, casting, forging, cutting, and grinding and polishing. Hiroshima prefecture

Table 6.1 Industrial manufacturing statistics (unit: billion Yen)

	Industry shipment amount			**Added value**		
	2002	2004	Growth rate	2002	2004	Growth rate
Nationwide total	260,259	275,726	1.06%	92,688	97,251	1.05%
01 Hokkaido cluster	5,043	4,976	0.99	1,661	1,677,721	1.01
02 Tohoku cluster	15,417	16,554	1.07	5,289	5,636	1.07
03 Kanto cluster	62,337	64,986	1.04	22,263	23,421	1.05
04 Hokuriku cluster	6,913	7,396	1.07	2,856	2,979	1.04
05 Chubu cluster	54,545	58,133	1.07	18,430	19,280	1.05
06 Tokai cluster	45,642	49,348	1.08	14,976	15,967	1.07
07 Kinki cluster	42,409	44,361	1.05	16,791	17,379	1.03
08 Chugoku cluster	19,364	21,318	1.1	6,529	6,982	1.07
09 Shikoku cluster	6,841	7,385	1.08	2,402	2,621	1.09
10 Kyushu cluster	17,501	18,682	1.07	6,072	6,500	1.07
11 Okinawa cluster	534	474	0.89	131	134	1.03

Notes

1: The constituent areas of the clusters are as follows. Hokkaido: Hokkaido. Tohoku: Aomori, Morioka, Iwate, Akita, miyagi, Fukushima. Kanto: Tokyo, Kanagawa, Saitama, Yamanashi, Gunma. Tokai: Aichi, Mie, Gifu. Chubu: Shizuoka, Aichi, Nagano. Hokuriku.

2: The prefectures listed are limited to those that have stated an industrial cluster plan.

3: Does not include biotechnology.

Source: Prepared by author based on METI's *"Kougyou Toukei"* (Industrial statistics).

is home to a variety of industries, including petrochemical, lumber, engine, automobile, electronics, shipbuilding, and machinery industries. Okayama prefecture includes electrical machinery, steel, fiber, automobile, and petrochemical industries. Yamaguchi prefecture has an agglomeration of oil-related firms, and there is a local initiative in the prefecture to form a high-technology cluster to produce the next generation of medical equipment.[8] This new cluster, centered in the Ube region of Yamaguchi prefecture, is working with area universities and corporate research institutions to develop and market medical light-source systems, low-invasive treatment equipment, and high-performance diagnostic equipment, through collaboration between medical and technical personnel.

Due to its history as a center of the automobile, shipbuilding, and industrial machinery industries, the Chugoku region possesses an outstanding pool of human resources, technology, and know-how, as well as a number of university-led projects promoted at the local level through collaboration between medical and technology and industry personnel.[9] We can thus assume that the factors driving the Chugoku region's cluster development predated the formation of the cluster; these factors include the existence of anchor firms and their related firms, and the existence of academic research institutions that collaborate with industry.

Development strategy and results so far

METI's project in this region aims to form an industrial cluster by developing new manufacturing businesses producing products for the world market, utilizing the full potential of local universities and research institutions in biotechnology, automobile manufacturing-related IT, shipbuilding, and industrial machinery.

The priority sectors listed in this project's vision statement are advanced medical equipment, human-assistance robotics, and next-generation mobile parts. Thus, support was devoted to those sectors. By 2005 this project had created 56 new local consortiums and 52 collaborative projects, investing 1.94 billion yen for technological development.

3.2.2. IT sector

METI is carrying out four projects involving IT clusters, located in the regions of Hokkaido, Kanto, Kinki, and Kyushu. We here focus on the Kyushu Silicon Cluster Plan in Kyushu, where growth has been high in recent years (Table 6.1).

Factors affecting cluster formation

Mitsubishi Electric laid the foundation for the Kyushu IT cluster when it built its semiconductor plant in Kumamoto Prefecture in 1967. Since then, major semiconductor manufacturers including NEC, Toshiba, Sony, Matsushita, Fujitsu and Oki Electric have built factories all over Kyushu.

Major manufacturers of semiconductor manufacturing equipment and inspection equipment have also opened plants in Kyushu. Agglomerations of local SMEs have formed in the areas around these semiconductor plants, consisting of manufacturers that mount semiconductors in integrated circuits as well as manufacturers of inspection equipment. In northern Kyushu, home to the cities of Fukuoka and Kitakyushu, an agglomeration of semiconductor design companies formed to take advantage of the research capabilities and human resources of Kyushu University. When IT sector clusters first formed in this area, there was already an agglomeration of a wide range of companies involved in semiconductor manufacturing, taking advantage of the fertile research environment. In this case, clustering efforts made optimal use of the technological potential and the special characteristics of Kyushu's existing semiconductor industry agglomeration, in a manner that enabled the creation of new businesses. Again, we find that the factors promoting cluster development pre-dated the cluster's formation, including the existence of anchor firms (here, large corporations) and related firms, and the presence of universities and researchers.

Development strategy and results

The semiconductor industry has been agglomerating in Kyushu for 30 years. This industry is capable of creating high-value-added products, and can be expected to boost industrial competitiveness across a wide spectrum. The Kyushu region is therefore promoting the Kyushu Silicon Cluster Plan with the aim of developing globally competitive companies and industries out from its current agglomeration. The Kyushu Semiconductor Innovation Council was established in 2002 to coordinate the implementation of this plan. The council encourages the creation of an environment conducive to innovation while also arranging collaborations among various actors in the outlying area.

So far, the plan has focused on developing next-generation products in the semiconductor sector, including intelligent appliances, robots, and LCD products. Currently, various programs are being implemented to help SMEs improve their technical capacity and reform their management. Venture companies are being promoted in order to accumulate know-how in the field of nanotechnology/MEMS (micro-eletromechanical systems), an applied technology that grew out of pre-existing semiconductor technology. The results of this plan as of 2005 consisted of the creation of 45 new local consortiums and 5 new technology products, with an investment of 1.59 billion yen for technology development.

3.2.3. Biotechnology sector

Four biotechnology cluster projects are underway in Japan, located in the regions of Hokkaido, Kanto, Tokai, and Kinki. Our discussion will focus on the Kinki Bio-Related Industry Project.

Factors affecting cluster formation

A number of major pharmaceutical companies originated in the Kinki region, and this area has long been home to a high concentration of pharmaceutical factories and related firms. High hopes have been placed on the Kinki region to develop biotechnology applications involving protein engineering, genetic engineering, and other cutting-edge techniques. Kinki is home to numerous public research institutions such as Riken (a physics and chemistry research institute), as well as universities that conduct advanced life science research, most notably Kyoto University and Osaka University. The Kinki region has thus possessed major pharmaceutical companies, who are the customers for biotechnology products, along with universities and research institutes that can help related firms respond to the anchor firms' needs.

This confluence of a core of universities with medical schools, pharmaceutical companies, and research institutes, surrounded by related firms, had a significant influence on cluster formation. Moreover, it appears likely that the active efforts of local government bodies were effective in promoting this cluster's formation.

Development strategy and results

The Kinki Bio-Related Industry Project is aimed at forming a globally competitive biotechnology cluster in Kinki with bio-venture companies at its core. The project is attempting to facilitate the creation of venture companies that utilize cutting-edge technology and to stimulate innovation, using biotechnology possessed by SMEs', thereby promoting the growth of export firms. This will be achieved by strengthening collaboration among industry, government, and academia, and through various measures that support technologies developed through university or enterprise R&D.

The project aims to strengthen competitiveness in its specified fields, with the aid of development strategies, particularly university-initiated ventures and collaboration between industry, academia, and government.[10] Bio-related ventures will form the cluster's core, benefiting from biotechnology transfers from SMEs that own high-tech capabilities, which will strengthen overall competitiveness in these specialized fields. Intensive investment in regenerative therapies and development of technologies related to the creation of pharmaceuticals will be key. Results of this project as of 2005 included the creation of 130 new local consortiums and the establishment of 15 collaboration programs, with the plan investing 4.15 billion yen for technological development.

3.2.4. Environmental technology sector

METI has four cluster projects in the environmental technology sector, located in the regions of Tohoku, Kinki, Chugoku, and Kyushu. Our discussion will focus on the Project for Promoting the Industry of a Recycling Society in the Tohoku region, where growth has been high in recent years.

Factors affecting cluster formation

The Tohoku region has long had numerous mines, and is a major production center of non-ferrous metals in Japan. Development and accumulation of advanced refining technology accompanied production activities. Many of the mines were forced to close down due to price wars with foreign producers and changes in demand, but there is now a movement in the region toward creating new environmental businesses by applying advanced technologies. One example of this is the application of refining technology developed over many years, primarily at Tohoku University, to separate and recycle useful metals from used household electrical appliances.

In addition, out of concern for the environment, the government is promoting "Eco-Town" projects. When Tohoku's clustering project was initiated, research on recycling technology and Eco-Towns[11] had already led to the creation of a number of related firms. In addition to human resources, technology, and know-how, the region had long possessed the infrastructure for collaboration among citizens, companies, and government, cultivated during the fight against pollution. Thus the factors affecting the formation of this local cluster appear to include the technology accumulated to support the mining industry and to control pollution, as well as the pre-existing agglomeration in the Tohoku region of research institutions and industry, especially energy and environment-related manufacturers.

Development strategy and results

METI's project in this region aims to form an environmental technology cluster by making use of core technologies developed in the Tohoku region, and to create and nurture new globally competitive industries to serve the industrial sector, where currently huge recycling needs are expected to expand even further in the future.

The project's strategy is to position the areas already designated as Eco-Towns as the core areas of the cluster. The strategy also includes creating businesses through collaboration among industry, academia, and government from seeds possessed by universities in the region, and initiating a technology development project to serve as a springboard for new businesses created through R&D carried out by regional consortiums. Results of this project as of 2005 included the creation of 39 new local consortiums and the establishment of 30 collaboration programs, with the plan investing 1.21 billion yen for technological development.

3.3. Hypothesis regarding the development of industrial clusters from industrial agglomerations

In all of the above case studies of cluster projects, the policies being implemented include three basic policy schemes.

The first policy scheme is network formation. Network formation projects may be broadly divided into two types: those that finance development of private promotional organizations that are involved in the formation of "face-to-face" collaborations, and those that finance collaborations with regional indust ry-support agencies in order to expand a cluster.

The second scheme is business support. Business support encompasses R&D support, sales route expansion support, business launching support, development of innovative institutions, support for collaboration among different industries, support for management reform, and support for human resources training.

The third scheme is promotion of tie-ups between firms and related agencies, including financial institutions, trading companies, educational institutions, etc.

Note that there are important elements that are common to all of the projects mentioned earlier, including the existence of anchor firms and corporate agglomerations in a specific sector, and the existence of a research infrastructure that specializes in that sector, that predates the cluster policy. As stated in Section 2, these anchor firms and corporate agglomerations were located in these regions in part due to the government's previous policy of relocating industry to outlying areas, embodied by the Technopolis Law, the Know-How Location Law, and the Local Base Law. With these elements already in place, existing universities and companies were then united by cooperation promotion agencies under the aegis of industrial cluster projects, leading to technological innovation.

Below in Figure 6.2 we present a flowchart model[12] of the formative factors and formative processes of industrial clusters in Japan, derived from the history of industrial cluster development and from case studies of the current projects. Based on this model, we next extract the formative factors through quantitative analysis of a questionnaire survey, as shown in the following section.

4. Research results and analysis of the factors underlying cluster formation and industrial cluster policy

To test our hypothesis regarding the development process of, and factors affecting, industrial cluster formation (Figure 6.2), we conducted a questionnaire survey (see Appendix for details) and performed quantitative analysis on the results.

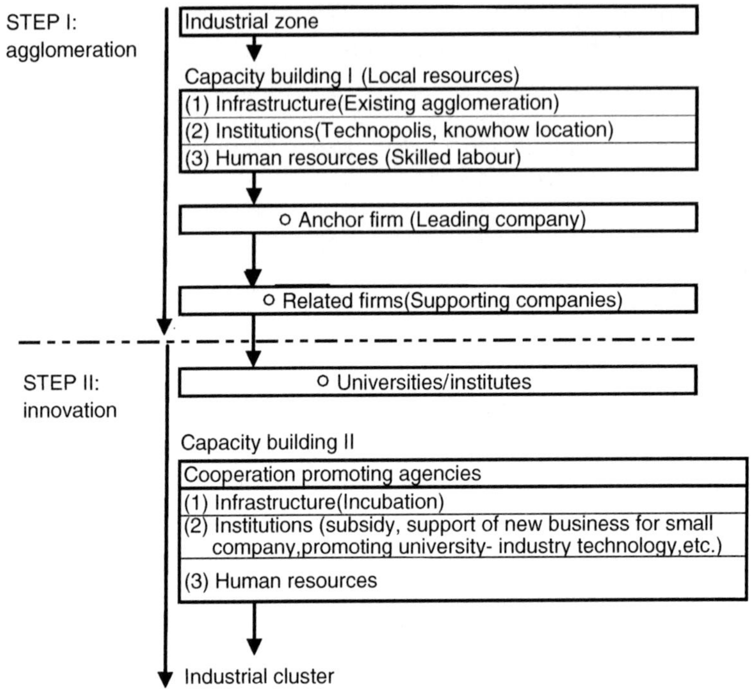

Figure 6.2 Development model of Japan's industrial clusters

Notes:

1. In this paper, I definite "Agglomeration theory is a mechanism for agglomeration, on the other hands Cluster theory is for a creation of an innovation " See. Ishikura., Y. ed."Nihon No Sangyo Kurasuta" 2003, Yuhikaku. Page 216.

2. STEP I is qoute by the prototype model of flowchart model in Chapter 1 (Kuchiki/ Tsuji), excluding "Living Condition."

Source: K. Yoshida and A. Kuchiki.

4.1. The formation of industrial clusters

4.1.1. Utilization of local resources

Some 87% of the METI agencies we approached for our interviews told us that their respective regions were taking advantage of local resources (Table 6.2). This usage of local resources is reflected in the following quotations from METI employees of descriptions of their areas:

- An industrial concentration in world-class manufacture of vehicles, ships and industrial machinery, supported by many companies in the machinery industry
- A concentration of enterprises related to the semiconductor industry

- A concentration of IT technology
- A concentration of research seeding, forming the basis for new kinds of manufacturing
- A concentration of small and medium size enterprises related to new kinds of manufacturing

There are many cases where pre-existing concentrations of big industries and of SMEs are being exploited to promote cluster development. There are also cases in which attempts are being made to cater to modern demand conditions by using the unique potential of particular locations. Examples of the latter, from METI responses, include "three eco-towns that are engaging in urban development using the special characteristics of the region. These are centered on industries related to recycling and the environment" and "developing the Kinki region's potential in life science fields." Half of our respondents who were active in cluster development are making use of infrastructure already developed under the "Technopolis" and "Knowledge Location" policies (Table 6.2).

4.1.2. Present awareness of industrial cluster formation

Some 56.8% of the local government offices that responded to our questionnaire survey said that industrial cluster formation was apparent in their areas. For JETRO[13] offices, this figure was 76.7% (Table 6.3). When asked

Table 6.2 Use of resources

Does the project about which you answered in No.1 have certain industry resources?

METI answer	Total	Yes	No
Total	23	20	3
	100%	87	13

Are previous industrial policies, such as Technopolis or Knowledge locations, still in effect?

METI answer	Total	Yes	No
Total	23	11	12
	100%	47.8	52.2

Source: author.

Table 6.3 Existence of industrial clusters

Do the firms in your prefecture currently form an industrial cluster?

(Local government)	**Total**	**Yes**	**No**
Total	37	21	16
	100%	56.8	43.2
(JETRO local Office)	**Total**	**Yes**	**No**
Total	30	23	7
	100%	76.7	23.3

Source: author.

what developments can be interpreted as an industrial cluster, answers varied from region to region, but at the national level, over half of respondents believed that an industrial cluster was some kind of industrial concentration associated with central government planning.

4.1.3. Preconditions for industrial cluster formation

Respondents were asked to identify local preconditions for industrial cluster formation in their areas. Some 95.7% of METI offices identified as preconditions the existence of related industries and of universities and research institutions, while 69.6% identified the presence of industries likely to form nuclei for growth. Local government offices also identified the existence of related industries and universities and research institutions (81.0% of respondents), universities and research institutions (71.4%), and the presence of industries likely to form nuclei for growth (47.6%). JETRO offices identified the presence of industries likely to form nuclei for growth (43.5%), the existence of related industries (34.8%), and the availability of high-quality human resources (30.4%) (Table 6.4).

It is clear from these replies that the existence of universities and research institutions of related industries, and of industries likely to form nuclei for growth were widely seen as important preconditions for cluster formation. Moreover, all three types of institution – METI offices, local government offices, and JETRO information centers – saw the existence of industries likely to form nuclei for growth as an important factor. For regions involved in the manufacture of vehicles, general machinery, electronic machinery, and software, the existence of related

Table 6.4 Factors for industrial clustering

What was the existing precondition for cluster formation?

METI answer	Total	Industrial zone	Nucleus firm*	Support of gov. policy	University/ institutes	Human resource pool	Environment	Related firm	Key person in the industry	Other
Total	23	6	16	9	22	15	4	21	2	2
	100%	26.1	69.6	39.1	95.7	65.2	17.4	91.3	8.7	8.7
Biotechnology/medical/ social welfare	10	2	6	5	10	6	1	8	1	1
	100%	20	60	50	100	60	10	80	10	10
IT/communications	5	2	3	1	5	3	2	4	0	0
	100%	40	60	20	100	60	40	80	0	0
Car assembly/machinery	6	2	3	1	6	3	1	5	0	0
	100%	33.3	50	16.7	100	50	16.7	83.3	0	0
Electronics devices/software	9	3	5	1	8	5	2	8	0	0
	100%	33.3	55.6	11.1	88.9	55.6	22.2	88.9	0	0
Liquid crystal display	4	2	3	0	3	2	1	4	0	0
	100%	50	75	0	75	50	25	100	0	0
Energy	2	0	1	1	2	1	1	2	0	0
	100%	0	50	50	100	50	50	100	0	0
Environmental recycling	6	1	2	4	6	3	1	5	0	1
	100%	16.7	33.3	66.7	100	50	16.7	83.3	0	16.7
Other	7	1	5	5	7	3	1	6	1	0
	100%	14.3	71.4	71.4	100	42.9	14.3	85.7	14.3	0

Continued

Table 6.4 Continued

Local government answer	Total	Industrial zone	Nucleus firm*	Support of gov. Policy	University/ institutes	Human resource pool	Environment	Related firm	Key person in the industry	Other
Total	21	7	10	6	15	8	4	17	4	3
	100%	33.3	47.6	28.6	71.4	38.1	19	81	19	14.3
Biotechnology/ medical	9	3	3	3	9	5	2	7	3	1
	100%	33.3	33.3	33.3	100	55.6	22.2	77.8	33.3	11.1
IT/communications	4	3	3	3	3	3	2	3		0
	100%	75	75	75	75	75	50	75	50	0
Car assembly/ machinery	10	3	6	1	5	4	2	9	1	2
	100%	30	60	10	50	40	20	90	10	20
Electronic devices	5	4	4	2	5	3	2	5	2	1
	100%	80	80	40	100	60	40	100	40	20
Liquid crystal displays	2	0	1	0	1	0	0	2	0	0
	100%	0	50	0	50	0	0	100	0	0
Other	6	1	3	2	6	2	1	6	1	0
	100%	16.7	50	33.3	100	33.3	16.7	100	16.7	0

JETRO answer	Total	Industrial zone	Anchor firm	Support of gov. policy	University/ institutes	Human resource pool	Environment	Related firm	Key person in the industry	Other
Total	23	3	10	6	6	7	6	8	1	4
	100%	13	43.5	26.1	26.1	30.4	26.1	34.8	4.3	17.4
Biology/medical	8	0	2	2	3	3	2	5	1	0
	100%	0	25	25	37.5	37.5	25	62.5	12.5	0
IT/communication	8	2	3	4	4	4	2	3	0	1
	100%	25	37.5	50	50	50	25	37.5	0	12.5
Car/machinery	6	1	3	1	0	2	3	3	1	1
	100%	16.7	50	16.7	0	33.3	50	50	16.7	16.7
Electronic device	8	2	5	1	1	3	1	3	0	2
	100%	25	62.5	12.5	12.5	37.5	12.5	37.5	0	25
Other	9	1	4	1	2	2	3	2	1	3
	100%	11.1	44.4	11.1	22.2	22.2	33.3	22.2	11.1	33.3

Summary of above 3 tables (Precondition)

METI, Local Government, JETRO	Total	Industrial zone	Nucleus firm*	Support of gov. policy	University/ institutes	Human resource pool	Environment	Related firm	Key person in the industry	Other
Total	67	16	36	21	43	30	14	46		9
	100%	23.9	53.7	31.3	64.7	44.8	20.9	68.7	10.4	13.4

* Firm likely to form a nucleus for growth.

Source: author.

industries and of industries likely to form nuclei for growth were seen as preconditions for cluster formation, whereas for biotechnology products, medical services, and products associated with public welfare, there was a strong tendency for the existence of universities and research institutions to be seen as a precondition. The following quotations from responses show why these various preconditions were identified by respondents.

First, regarding the importance of the existence of related enterprises, including anchor firms, respondents said:

- In photovoltaic cell production, the big companies concentrated in the Kinki region account for over 40% of world output.
- In fuel cell and hydrogen production, leading manufacturers of solid high polymer-type fuel batteries for domestic co-generation equipment are already present, as are important manufacturers of hydrogen, and there are facilities for transport, storage, and conservation.
- Many makers of biomass energy production plants are already present.
- So far as second-generation battery production is concerned, not only are the major manufacturers already present [here], but there are also universities and research institutions with strong electrochemistry departments.
- Companies with good competence for development in basic and peripheral technology are present.
- In the environment-related field, the leading makers of environment-related equipment are present.

These and similar observations show that the presence in the area of leading companies and top manufacturers is perceived as important for cluster formation. Officials reason that "because of the existence of companies likely to form nuclei for growth, an early concentration of related industries is a possibility," and "for promoting the exchange of information, and for regulating the possibly conflicting interests of firms, it is essential that there already exists a 'nucleus' company with negotiating power based on the scale of operations and technology."

Among the replies, there were frequent references to the importance of the presence of companies likely to form nuclei for growth. Quotations reflecting this view include:

- Sankyo Seiki Seisakusho was formed as a 'spin-off' company from Seiko-Epson.
- Big companies already present, such as Toshiba, have functioned as magnets for the formation nearby of concentrations of small and medium-sized enterprises (locally-based industries), and have thus fostered the concentration of human resources.

- The existence of big companies has a strong influence on the formation of industrial clusters. The existence of one large firm can have a big impact on the local economy.

The following quotations are also significant:

- In the Project to Create New Manufacturing in Hokuriku, three workshops were set up: the Hokuriku Life Care Cluster workshop, the Hokuriku Micro-Nano Process workshop, and the Hokuriku Advanced Complex Materials workshop. Of these workshops, two have become energetic offices of key companies in the cluster.
- By linking with the concrete results of vigorous R&D... companies likely to form nuclei for growth have acted as coordinators for the activities of related enterprises, a feature that is highly effective for the formation of an industrial cluster.

Respondents also pointed to the importance of universities and research institutions:

"In the life science field, world-class centers of excellence such as Kyoto University, Osaka University, the Center for Developmental Biology, the National Institute of Advanced Industrial Science, and the Kansai Center of the National Institute of Advanced Industrial Science and Technology (AIST) are all located in the region. Furthermore, companies that have considerable technological and financial power are locally present in the pharmaceutical industry (in the form of leading manufacturers) and in industries such as chemicals and foods, while there are also small and medium size enterprises equipped with great technological competence. All this makes it possible to bring about industrial development that combines biotechnology as a core technology with other peripheral technologies."

On the importance of universities as places for fostering entrepreneurial activity, one respondent said:

> The activities of Hokkaido University's 'Microcomputer Workshop' provided the initiative that led to the beginnings of venture capital businesses in information-related fields such as software and system house development, and these led to the formation of Sapporo's concentration of information industries. In this way, the existence of Hokkaido University was an important factor in the formation of the cluster.

Other responses illustrate the important contribution made by collaboration between university research laboratories and companies, especially with regard to business initiatives and technology:

- It is apparent that there is a movement, aimed at making use of the research seeds sown by local universities and research institutions, to develop new enterprises with technology suitable for raising the added value of production.

- Many companies have moved into the area to make use of the facilities provided by Nihon University's Worldwide Research Center for Advanced Engineering and Technology (NEWCAT) in Koriyama.
- As part of Kobe's Medical Industry City Plan, thanks to initiatives taken by the local government, 76 private research laboratories are now concentrated within the city.

Other opinions were put forward by smaller numbers of respondents. One response showed that there have been cases in which links with financial institutions have helped to encourage growth:

> In the region, local financial institutions have created links with facilities such as the Technology Advanced Metropolitan Area (TAMA) foundation. This has had important implications for cluster development and has contributed to the growth of companies within the cluster. It is hoped that this situation will be replicated in cluster areas elsewhere in Japan.

Another respondent offered the following observation:

> A fundamental precondition for cluster formation is the presence within the region of existing companies and universities. But even more important is the local presence of a leader (someone with charisma). In Mie prefecture, this role is played by someone whom I will call 'Mr M.' For ten years, Mr M. has worked enthusiastically on the plan for the formation of Medical Valley in the prefecture. Important contributors such as industry associations and Mie University have all relied on his ideas. Without him, there would have been no Medical Valley – and the Valley has been ranked the fifth most important bio-cluster in Japan.

4.1.4 Institutional support for the formation of industrial clusters

The availability of supportive subsidies and loans was identified as the most effective government cluster policy by all three types of respondents, namely METI (100%), local government offices (90.5%), and JETRO centers (60.6 %.). METI was identified as the most important single policy,that is the New Business Promotion Act for Small Firms (47.8% of respondents), while local government and JETRO were singled out as the next most important policies,that is the Act for the Promoting the Relocation of University Technology and the legislation for Designated Structural Reform Districts (Table 6.5). Some replies also mentioned as important, subsidies

Table 6.5 Effective government policy

What is the government policy or policies that is/are effective to form an industrial cluster?

METI answer	Total	Law for promoting university-Industry technology	Promotion of special zones for structural reform	Financial aid	Revised industrial revitalization law	Promotion of small business establishments	Other
TOTAL	23	7	0	23	0	11	3
	100%	30.4	0	100	0	47.8	13
Biotechnology/medical/ social welfare	10	4	0	10	0	3	1
	100%	40	0	100	0	30	10
IT/communications	5	2	0	5	0	2	1
	100%	40	0	100	0	40	20
Car assembly/machinery	6	2	0	6	0	3	0
	100%	33.3	0	100	0	50	0

Continued

Table 6.5 Continued

METI answer	Total	Law for promoting university-industry technology	Promotion of special zones for structural reform	Financial aid	Revised industrial revitalization law	Promotion of small business establishments	Other
Electronics devices/software	9	3	0	9	0	5	1
	100%	33.3	0	100	0	55.6	11.1
Liquid crystal displays	4	1	0	4	0	1	0
	100%	25	0	100	0	25	0
Energy	2	0	0	2	0	2	0
	100%	0	0	100	0	100	0
Environmental recycling	6	0	0	6	0	2	1
	100%	0	0	100	0	33.3	16.7
Other	7	3	0	7	0	3	1
	100%	42.9	0	100	0	42.9	14.3

Local government	Total	Law for promoting university-industry technology	Promotion of special zones for structural reform	Financial aid	Revised industrial revitalization law	Other	N.A.
Total	21	4	3	19	1	0	2
	100%	19	14.3	90.5	4.8	0	9.5
Biotechnology/medical	9	1	2	9	0	0	0
	100%	11.1	22.2	100	0	0	0
IT/communications	4	0	1	4	0	0	0
	100%	0	25	100	0	0	0
Car assembly/machinery	10	2	1	8	0	0	2
	100%	20	10	80	0	0	20
Electronic devices	5	0	1	5	0	0	0
	100%	0	20	100	0	0	0
Liquid crystal displays	2	1	0	2	0	0	0
	100%	50	0	100	0	0	0
Other	6	3	1	6	1	0	0
	100%	50	16.7	100	16.7	0	0

Continued

Table 6.5 Continued

JETRO	Total	Law for promoting university-industry technology	Promotion of special zones for structural reform	Financial aid	Revised industrial revitalization law	Other	N.A.
Total	23	4	4	16	0	2	4
	100%	17.4	17.4	69.6	0	8.7	17.4
Industrial zone	3	0	0	2	0	0	1
	100%	0	0	66.7	0	0	33.3
Nucleus firm	10	1	1	6	0	1	2
	100%	10	10	60	0	10	20
Support of gov. policy	6	2	2	6	0	1	0
	100%	33.3	33.3	100	0	16.7	0
University/institutes	6	2	2	5	0	1	0
	100%	33.3	33.3	83.3	0	16.7	0
Human resource pool	7	0	1	6	0	1	1
	100%	0	14.3	85.7	0	14.3	14.3
Environment	6	2	2	5	0	2	1
	100%	33.3	33.3	83.3	0	33.3	16.7
Related firm	8	1	2	5	0	1	1
	100%	12.5	25	62.5	0	12.5	12.5
Key person in the industry	1	0	0	0	0	0	1
	100%	0	0	0	0	0	100
Other	4	0	0	2	0	0	2
	100%	0	0	50	0	0	50

Source: author.

for helping firms grow by supporting technology, and funds for promoting consortiums.

87.0% of METI, 52.4% of local government, and 30.4% of JETRO responses cited the importance of supportive policies for the formation of clusters. One reply said: "A comprehensive mechanism that takes marketing strategy into account needs to be created." Other similar opinions referred to the need for "supporting policies for marketing research and distribution." Additionally, responses mentioned that "human resource management is needed to organize an industry cluster project," "human resource development is needed for the future [for both company management and intellectual property strategy]," and "there is a need for the development of coordinators." Many areas where industry cluster projects are already underway now urgently need policies supporting marketing, strategic planning, and human resource management.

Moreover, "companies have sought supportive policies for new industries," such as "extension of the Green Procurement Project [generating demand for recycled products]," "support for the commercialization of the bio-venture business," or "policies assisting venture companies [loan prioritization and coordination]." One respondent felt that there was a need for a bold policy "to ease regulations in special financial business zones."

Some 28.6% of METI local offices believe that there is a demand for the adjustment of policies toward supporting the evolution of industry clusters; 14.3% of local government offices and 13.0% of JETRO trade information centers shared this view. "Establishment of a scenario" and "research and development by public agencies" were perceived as necessary starting points for the implementation of government long-term planning. Also, it was felt that there is a need for "skill angels," meaning business-supportive staff from local governments, as well as an "improvement of the government's coordinating function."

4.1.5. The role of cooperation-promoting agencies in industrial cluster formation

The three types of government agencies surveyed for this study have different functions, but often find themselves in similar positions when it comes to encouraging industrial cluster formation. For example, all three agencies seek to set up of networks of collaboration among industry, government, and academia, as well as to implement initiatives designed specifically for small companies, as well as the export promotion. When asked what would be the best way to encourage the creation of new technologies, new business enterprises, and new industries, METI agencies chose the promotion of collaboration between business and academia, and networking and matching assistance (91.3% each), while local government offices selected the promotion of collaboration between business and academia (95.2%) and specific support for small firms (39.1%). JETRO picked export promotion (56.5%), the attraction of enterprises (43.5%), and industry-government collaboration (39.1 %) (Table 6.6).

Table 6.6 Role of coöperative promotion agencies

What activities should the government engage in to develop industrial clusters?

METI answer	Total	Attraction of enterprises	Improvement of infrastructure	Policies for small businesses	Policy support
Total	23	2	0	5	4
	100%	8.7	0	21.7	17.4
Biotechnology/ medical/ social welfare	10	1	0	0	2
	100%	10	0	0	20
IT/communications	5	1	0	2	2
	100%	20	0	40	40
Car assembly/ machinery	6	0	0	1	0
	100%	0	0	16.7	0
Electronics devices/ software	9	1	0	3	2
	100%	11.1	0	33.3	22.2
Liquid crystal displays	4	0	0	1	1
	100%	0	0	25	25
Energy	2	0	0	1	0
	100%	0	0	50	0
Environmental recycling	6	1	0	1	0
	100%	16.7	0	16.7	0
Other	7	2	0	2	1
	100%	28.6	0	28.6	14.3

4.1.6. *Competitive advantage of areas where clusters are being formed*

When asked how a region can build competitive advantage in the sectors in which it possesses incipient industrial clusters, METI stressed the development of new technology and the creation of a human resource pool (52.2%). Local government offices identified new technology (85.7%) and the creation of a human resource pool (23.8%), and the JETRO centers opted for the development of new technology (30.4%) and the creation of a human resource pool (26.1%) (Table 6.7).

Export promotion	Industry-gov.-academia corporation	Network-building	Provision of information	Matching assistance	Overseas operation backup	Other
0	21	21	19	21	10	3
0	91.3	91.3	82.6	91.3	43.5	13
0	10	9	8	9	5	1
0	100	90	80	90	50	10
0	4	4	4	4	3	1
0	80	80	80	80	60	20
0	6	4	3	5	2	2
0	100	66.7	50	83.3	33.3	33.3
0	8	7	8	8	4	2
0	88.9	77.8	88.9	88.9	44.4	22.2
0	3	3	3	3	1	0
0	75	75	75	75	25	0
0	2	2	2	2	0	0
0	100	100	100	100	0	0
0	5	6	5	5	1	0
0	83.3	100	83.3	83.3	16.7	0
0	7	7	6	7	2	2
0	100	100	85.7	100	28.6	28.6

Continued

Table 6.6 Continued

Local government answer	**Total**	**Attraction of enterprises**	**Improvement of Infrastructure**	**Policies for small businesses**
Total	21	3	3	6
	100	14.3	14.3	28.6
Biology/medical	9	1	2	1
	100	11.1	22.2	11.1
IT/communication	4	2	2	1
	100	50	50	25
Car/machinery	10	1	3	3
	100	10	30	30
Electronic device	5	1	2	1
	100	20	40	20
Liquid crystal display	2	0	1	0
	100	0	50	0
Other	6	1	0	2
	100	16.7	0	33.3

JETRO Answer	**Total**	**Attraction of enterprises**	**Improvement of infrastructure**	**Policies for small businesses**
TotaL	23	10	6	6
	100%	43.5	26.1	26.1
Biotechnology/medical	8	4	3	2
	100%	50	37.5	25
IT/communications	8	3	3	1
	100%	37.5	37.5	12.5
Car assembly/machinery	6	1	3	2
	100%	16.7	50	33.3
Electronic devices	8	2	2	2
	100%	25	25	25
Other	9	4	1	2
	100%	44.4	11.1	22.2

Source: author.

Policy support	Export promotion	Industry-gov.-academia corporation	Other	N.A.
5	1	20	3	0
23.8	4.8	95.2	14.3	0
2	1	9	0	0
22.2	11.1	100	0	0
0	0	4	1	0
0	0	100	25	0
2	0	9	1	0
20	0	90	10	0
1	0	5	1	0
20	0	100	20	0
1	0	2	0	0
50	0	100	0	0
3	1	6	1	0
50	16.7	100	16.7	0

Policy support	Export Promotion	Industry-gov.-academia collaboration	Other
6	13	9	4
26.1	56.5	39.1	17.4
2	5	3	1
25	62.5	37.5	12.5
1	5	4	0
12.5	62.5	50	0
1	2	3	1
16.7	33.3	50	16.7
2	5	2	1
25	62.5	25	12.5
1	6	2	1
11.1	66.7	22.2	11.1

Table 6.7 Advantages of industrial custers

What is the major advantage of your industry cluster being in the region concerned?

METI answer	Total	Technology innovation	Human resource pool	Information network	Increasing productivity	Beneficial institutions	N.A.
Total	23	12	12	8	3	2	5
	100%	52.2	52.2	34.8	13	8.7	21.7
Biotechnology/medical/social welfare	10	7	4	3	2	2	2
	100%	70	40	30	20	20	20
IT/communications	5	2	3	2	0	1	1
	100%	40	60	40	0	20	20
Car assembly/machinery	6	4	2	1	0	0	2
	100%	66.7	33.3	16.7	0	0	33.3

Electronics devices/software	9	5	4	5	0	1	1
	100%	55.6	44.4	55.6	0	11.1	11.1
Liquid crystal displays	4	2	2	2	0	0	1
	100%	50	50	50	0	0	25
Energy	2	1	1	1	0	0	1
	100%	50	50	50	0	0	50
Environmental recycling	6	3	3	1	0	1	1
	100%	50	50	16.7	0	16.7	16.7
Other	7	2	3	1	1	1	2
	100%	28.6	42.9	14.3	14.3	14.3	28.6

Continued

Table 6.7 Continued

Local government answer	Total	Technology innovation	Human resource pool	Shared information	Increasing productivity	Beneficial institutions	N.A.
Total	21	18	5	4	3	3	1
	100%	85.7	23.8	19	14.3	14.3	4.8
Biotechnology/medical	9	9	4	1	1	2	0
	100%	100	44.4	11.1	11.1	22.2	0
IT/communications	4	4	3	1	0	0	0
	100%	100	75	25	0	0	0
Car assembly/machinery	10	9	3	3	3	2	0
	100%	90	30	30	30	20	0
Electronic devices	5	4	4	1	0	0	1
	100%	80	80	20	0	0	20
Liquid crystal displays	2	2	0	1	2	2	0
	100%	100	0	50	100	100	0
Other	6	4	0	1	1	1	1
	100%	66.7	0	16.7	16.7	16.7	16.7

JETRO answer	Total	Technology innovation	Human resource pool	Shared information	Increasing productivity	Beneficial institutions	N.A.
Total	23	7	6	1	1	5	6
	100%	30.4	26.1	4.3	4.3	21.7	26.1
Law for promoting university-industry technology	4	2	0	0	0	1	1
	100%	50	0	0	0	25	25
Promotion of special zones for structural reform	4	1	0	0	0	2	1
	100%	25	0	0	0	50	25
Financial aid	16	4	5	1	1	5	3
	100%	25	31.3	6.3	6.3	31.3	18.8
Other	2	1	1	0	0	1	0
	100%	50	50	0	0	50	0
N.A.	4	1	1	0	0	0	2
	100%	25	25	0	0	0	50

Source: author.

4.2. The Future of industrial cluster formation

4.2.1. Areas already developing industrial clusters

When asked what future approaches should be taken with regard to industrial cluster formation in Japan, respondents wrote:

- Hitherto, we have focused on services for networking and technology development only for member companies, and for this reason a distinctive industry cluster has not yet emerged in the manufacturing sector. Our next goal will be to provide services for forming a distinctive industrial cluster.
- To promote innovation as the main objective of industry cluster projects, I would like to emphasize the creation of networks among various industries. It will also be essential to find and where necessary educate coordinators and advisers for developing such networks.

These opinions indicate that the optimal role for policymakers changes when industrial cluster formation proceeds past the initial stage. Whereas in the initial phase a formal step-by-step approach is necessary, in the subsequent stage development should be much more tailored to local needs, with member companies sharing a clearly defined mission statement and engaging in substantial collaboration toward mutual goals. In this context, the demand for coordinators will grow as clusters move toward high-level collaboration. Another respondent said:

> In the semiconductor field in the Kyushu region, we have held many different kinds of seminars, and have undertaken various initiatives for promoting technological development and sales development. Since starting the cluster plan four years ago, we have achieved substantial results. The number of member companies involved with the promotion agencies has increased to 1,119 as of August 2005. We have also established a network linking industry, academia, and government. Since the promotion act for Kyushu's semiconductor industry was introduced, the industry has been efficiently activated. We would like to reinforce our local networks with the newly established 'Wide Area New Business Network Reinforcement Project.

It is apparent that some areas where cluster projects have been developed are trying to expand the range of fields in which collaboration takes place. Many of these inter-category collaborations have occurred in areas that have been trying to engage in industry cluster projects for a relatively long time.

Many respondents mentioned goals for the next stage of cluster formation:

> We are currently working on joint research and information exchange to develop a healthy long life industry and new energy source industries. So

> far, we have concentrated on traditional industries and on modern measurement analysis and electronic devices. Now, we plan to promote new wellness-related enterprises or a trial manufacturing industry cluster incorporating horizontal technology development linked with local technology.

These are the roles that respondents expect the government to fill in the formation of industry clusters. However, it is also hoped that private firms will be capable of forming clusters on their own:

> A president of a medical device company in Lombardia, Italy, opened a new company in a small village. The firm was growing rapidly, and he sold the company and set up a related company near the original one. Finally, there were more than eighty medical-related firms agglomerated in and around the tiny village. However, the village attracted little attention from the cities because the government [policy] could not keep up with the speed of [the cluster's] growth. Therefore, what the government should do is to restrict itself to supporting private companies trying to create clusters.

4.2.2. Areas with no current industry cluster

Local government offices that do not at present have an industry cluster present in their territories wish to attract firms capable of acting as anchor firms and forming growth nuclei for future clusters (40.0%). These offices also desire involvement by universities and research institutes (33.3%), as well as the presence of charismatic leadership (26.7%). JETRO also desires universities and research institutes, firms that are related to one another, and human resources (42.9% each) (Table 6.8).

We can see from these responses that the elements most desired for cluster formation are firms capable of forming growth nuclei, related firms, universities and research institutes, and a pool of quality human resources. Some specific responses on this topic included:

- From 2002 to 2004, we repeatedly discussed a plan for creating an industry cluster in our area, and I felt there must be an clear leader to act as a coordinator. – The demand for a leader to organize a concentration of related firms is clearly apparent here, as is the need for consortiums.
- Two regions have been selected as "Ministry of Education and Science City Area Business Projects," thanks to the seeds generated by new technologies. Although these regions are currently approaching the stage where they will develop into "Intellectual Clusters," they will need an "area consortium research and development project" to link the intellectual clusters with an industry cluster...defined as the creation of new businesses out of a network involving industry, academia and government.

- Several kinds of seeds [have the potential to] generate an industry cluster; however, a real industry cluster, which does not have functional characteristics that clusters are supposed to have, has yet not appeared. The reason might have something to do with the existence of a business concentration in Tokyo, which means that there is an outflow [of firms and talent] from local sources to the Tokyo area.
- Competition among cities beyond [Japan's] national boundaries has begun, and if this trend continues, international economic barriers may eventually be eliminated. Until Japan finds a clear solution to that problem, it will be difficult to create industrial clusters in Japanese provinces.

4.3. Analysis of results

Currently, the most common kinds of industrial cluster in Japan are those with industries based on biotechnology and medicine, social welfare, electronic devices, and software. These clusters often incorporate existing large companies or an agglomeration of small firms, and depend on these enterprises for cluster development. It is clear that there are several cases where it has proven possible to match locally-specific technology potential with current market demand. These successes include clusters described by our respondents as "an Eco-town project with the environment-related and recycling industry as its core element" and "high technology potential in

Table 6.8 Future plans for forming industrial clusters

What is the most important factor likely to encourage the future growth of an industrial cluster in your area?

Local government answer	**Total**	**Industrial zone**	**University/ institutes**	**Related firm**
Total	16	0	6	3
	100%	0	37.5	18.8

JETRO answer	**Total**	**Industrial zone**	**University/ institutes**	**Related firm**
Total	7	0	3	3
	100%	0	42.9	42.9

Source: author.

the field of the life-science industry." In the next paragraph, the process and fundamental stages of industrial cluster formation will be examined.

The exploitation of local resources has played an important role in cluster formation. Here, local resources include natural resources ranging from plant and water resources (whose endowment reflects specific climate and other geographic conditions) to energy resources, such as solar power or fuel batteries. Broadly defined, "local resources" can also include the presence of existing industrial agglomerations, small manufacturing companies, and skilled technicians in an area. These factors comprise the (1) infrastructure, (2) institutions, and (3) human resources elements of the capacity building part of the flowchart approach. In Japan, these local resources often include infrastructure and skilled laborers involved with the industrial estates formed through the "technopolis" and "know-how" policies of the past.

Japan's industrial clusters have formed by utilizing existing resources such as these. Half of the clusters we studied made use of infrastructure that was put in place by earlier industrial policies (technopolis[14] and know-how locations[15]). However, it must also be pointed out that approximately half of the industrial agglomerations formed spontaneously, without relation to any industrial policy, by firms making use of local resources (such as natural resources) that existed in their areas.

Our surveys demonstrated that the major preconditions needed for the formation of an industrial cluster are the presence of related firms, firms likely to function as nuclei for growth, and universities and research

Nucleus Firm	Human Resource Pool	Key person in the industry	Support of Gov. Policy	Environment	Other	N.A
6	1	4	0	1	3	1
37.5	6.3	25	0	6.3	18.8	63

Nucleus Firm	Human Resource Pool	Key person in the industry	Support of Gov. Policy	Environment	Other
1	3	0	0	0	1
14.3	42.9	0	0	0	14.3

institutes. In particular, the existence of firms likely to function as nuclei for growth is the most important basic requirement for cluster formation. In vehicle assembly and machinery production, in electronic device assembly, and in the software industry, firms likely to function as nuclei for growth and related firms are the most important requirements; however, the growth of biotechnology, medical service industries, and social welfare industries has tended to be associated with the presence of universities and institutes.

The growth factors differ from category to category because, as one respondent put it, "in an industry agglomeration area, the existence of a big company, which can be a firm likely to function as a nucleus for growth, and one with high development capacity, makes it far easier for other firms to form an agglomeration in the early stages of cluster formation," while "the strength of big companies in terms of capital, [combined with] the [specialized] high technology of smaller firms, can enable the joint development of industry by combining core technologies with peripheral technologies." Moreover, by making use of seeds provided by universities and research institutes, industries can generate new technology as well as businesses with highvalue-added production. The collaboration of firms with universities and research institutes can contribute substantially to the creation of new technologies and new types of business. We can thus conclude that agglomerations will be created by mutually supportive industries centering on firms functioning as growth nuclei, with related firms and new incoming firms following the lead taken by big companies. The involvement of academic entities such as universities and research institutes with the nuclei firms and related firms in an agglomeration signals the evolution of the agglomeration into an industrial cluster, which is necessarily an environment of creative innovation.

Cash subsidies and financial support policies were the approaches most highly regarded by our respondents. Another popular approach was the "New Business Supporting Act" for smaller firms- a means of providing direct support for smaller companies. Some respondents approved of the "Act for Supporting the Relocation of University Technology," which provides patent protection and which recognizes the importance of technology relocation to private companies. It follows that in the early stages of industrial cluster development, the most strongly desired support policies are those that directly assist small companies and that reinforce extensive cooperation between related firms and universities and research institutes. Patent protection and technology transfer, both of which can be promoted by collaboration between universities and private companies, are also highly regarded as an approach to cluster development.

Four years have passed since the government introduced its industrial cluster policy, and by now each region in Japan has at least one new challenge to tackle in order to improve their clusters. In our survey, more than 90% of METI branches thought that current support policies ought to be adapted to better suit local circumstances, and more than half of local government offices and 30% of JETRO trade information centers agreed.

The results of our survey suggest that a desirable new support policy would be one that gives attention to marketing strategies for product commercialization, as well as to marketing research. Another useful policy would be one that recognizes the importance of human capital development for the management of industrial cluster projects, for creating new businesses, and for coordinating the activities of the main actors in a cluster.

Some of our respondents also felt that there was a need for policies to support the development of venture capital businesses. Each of the three agencies whose views we investigated agreed that there was a need for support for individuals able to advise companies on marketing, especially companies in new and growing industries. They also spoke of the need for specialists able to coordinate the main trends within a cluster and organize these with a clear vision in mind. Thus far, in the absence of individuals with charismatic leadership qualities, these roles have been played by industrial promotion agencies and by collaborations among industry, government, and academia. There is a clear demand for agencies and people with leadership qualities to encourage and organize collaboration and regional cooperation among these various actors.

We ask: What have been the benefits of industrial clusters, and what are government's future tasks regarding industrial cluster formation? Judging from the replies we received, many regions have benefited from the growth of industrial clusters, especially in cases where engineering and R&D have created marketable new products. The biggest single benefit provided by industrial clusters has been product innovation, which is different from the process innovation that had been predominant in Japanese firms until recently. Regional clusters are increasingly owning new technologies and valuable human resources, which can be inherited by the next generation.

In areas that presently have no industry cluster, the local offices of METI, local governments, and JETRO believe that the driving force for future industrial cluster formation will come from firms likely to provide growth nuclei, from universities and research institutes, and from human capital resources. Areas without a cluster but wishing to create one should try to identify "regional resources" that can generate "added value" by using "technology development and transferred technology created by collaboration among universities, research institutes, and other actors in a cluster."

The major player within an industry cluster is the private firm. Government needs to encourage the formation of "region-oriented" industrial clusters, and should support the creation of an environment in which companies can gain competitive strength. Firms, for their part, should maximize usage of their local advantages to both compete and cooperate constructively.

5. Conclusion

Although Japan's current industrial cluster policy has had a significantly positive impact on the country's economy, it has several weaknesses. Based on the results of our study, we made the following observations regarding Japan's industrial cluster policy.

In the initial stage of cluster formation, pre-existing industrial agglomerations of large companies and SMEs are often utilized as resources. In each such region, market demand was effectively matched in some way with a region's particular potential. Drawing on the base of pre-existing industrial agglomerations (sometimes formed by past industrial policies), the current industrial cluster policy is effective as a tool for spurring innovation (Figure 6.3: Step 1, Capacity Building).

In an industrial cluster, it is necessary to have both anchor firms and related firms. The most vital factor for cluster formation is the existence of anchor firms. Leading companies and top manufacturers, which have strong development capacities in the form of both infrastructure technology and peripheral technology, can act as anchor firms, enabling the rapid agglomeration of related firms and, eventually, the formation of industrial clusters (Figure 6.3: Step 1, Anchor Firms, Related Firms).

Furthermore, the presence of universities and research institutes is indispensable to the formation of industrial clusters. These institutions make tremendous contributions to technology and business development by collaborating with private firms (Figure 6.3: Step 2, Universities/Institutes).

In new knowledge-intensive industries, old-fashioned agglomeration strategies must give way to innovation-intensive strategies. This new focus is evident in the infrastructure building (incubation), system support, and human resources training being conducted as part of industrial cluster policies across Japan (Figure 6.3: Step 2, Capacity Building). Direct grants and loans to SMEs have proven to be effective in this regard, but as yet have had no effect on living environments in clusters.

The presence of one or more key individuals (charismatic leaders) is very significant for a cluster's success. Also, collaborations between various parties in an area and utilization of partnerships are critical. Currently, cooperation promotion agencies are attempting to fill leadership roles in Japan's industrial cluster projects (Figure 6.3: Step 3, Capacity Building).

Each clustering strategy and project involves a wide variety of industries other than high technology industries, and policies are often directed toward existing industries and their agglomerations. We believe that it is effective to invest new knowledge and technology in existing industries and their agglomerations, and to step up the revitalization of these industries.

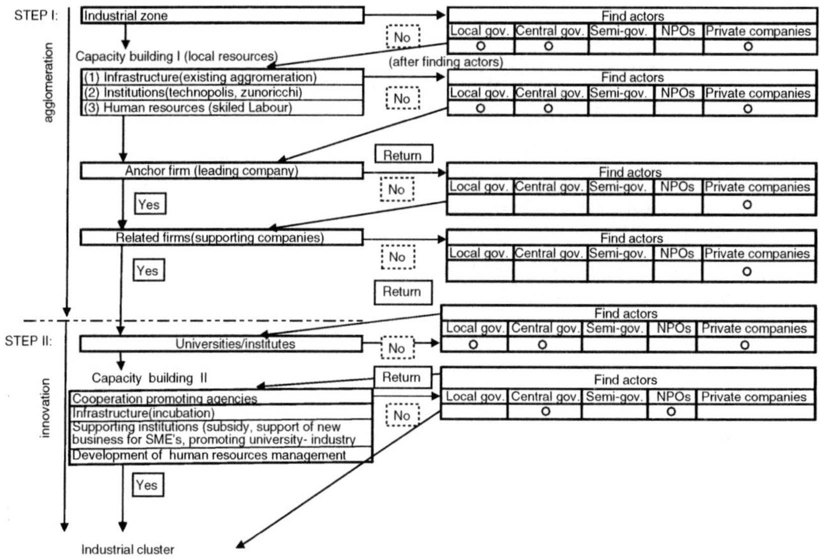

Figure 6.3 Flowchart model driven from the research

Notes:

1. The listed factors are effective assumptions for a formation of industrial clusters quantitatively proved by this survey.

2. Infrastructure in Capacity Building II was not proved in this survey; however, it is implemented by actual policies.

3. Development of Human resources management in Capacity Building was not proved in this survey; however, it exists as an actual political scheme. Additionally, it remains as an issue to besolved as comments in the suervey.

Source: authors.

We make the following conclusions:

The presence of the "flowchart elements" of anchor firms, related firms, and capacity building in industrial clusters (which are by definition environments of innovation creation), was quantitatively demonstrated for the case of Japan's clusters.In Japan's case, 1) infrastructure development (incubation), 2) system support for institutions, and 3) human resources development are the essential elements of Capacity Building II in our flowchart model.

There are numerous regions creating innovations even without the presence of a charismatic leader. However, these regions have not produced world-renowned "super clusters" like Silicon Valley or Austin.

Finally, the presence of cooperation promotion agencies is necessary to drive the process of innovation that characterizes industrial clusters.

The study also identified the following limitations of Japan's current cluster policy:

Despite the high priority the Japanese government has placed on cluster policy, the results of these policies so far have been moderate, and world-class clusters along the lines of Silicon Valley and Austin have so far failed to emerge.

Specialized marketing knowledge is now needed to "popularize" the new products produced in the clusters. Creating new markets, which means developing customers in addition to products, is distinct from network formation and should be treated as such.

In Japan's clusters, progress has been limited in encouraging the active participation of SMEs, in collaboration between companies, and in cluster-supporting collaborations between firms and the government. Moreover, even where such collaborations are being carried out, almost no charismatic individuals have emerged to give clear vision and leadership to the clusters.

Given these limitations and difficulties, we believe that follow-up studies should be conducted. The relationship between development factors and the development process should be quantitatively verified. Whether or not the presence of an anchor person (charismatic individual) is a pre-requisite for a strong cluster should also be quantitatively verified.

Appendix

An investigation of the factors responsible for the formation of industrial clusters, and of industrial cluster policy: outline of the study.

1.1. Research aims

In 2005, a research group was set up by the Institute of Developing Economies with the title "Comparisons of Industrial Concentrations in Asia and Other Regions – the Factors Responsible for the Development of Concentrations." The research presented in this chapter comprises part of that study, examining policies towards industrial clusters in Japan and the factors responsible for the formation of such clusters. Our research examined the main constituents of policy in Japan's regions and investigated the institutions responsible for the implementation of industrial cluster policy. We also attempted to determine which policies will be most effective for future development. The authors gathered basic information in this study, with the ultimate intention of carrying out international comparisons. It is hoped that the findings of this project will provide a basis for identifying more effective cluster support policies.

1.2. Institutions studied

In 2001, the Ministry of Economy Trade and Industry (METI) announced its industrial cluster policy, which forms a central part of industrial policy in Japan. There are two kinds of industrial cluster policy in Japan, one being METI's "Industrial Cluster Plan" (introduced in 2001) and the other being the "Project for Creating Knowledge Clusters" inaugurated in 2002 by the Ministry of Education, Culture, Sports, Science and Technology. Under METI's initiative, these policies provide clusters with support by encouraging collaboration on a variety of levels. Local governments are also involved in industrial cluster formation, working in line with national industrial cluster policy. The present research investigated the actions taken by publicly-administered economic and industrial agencies throughout Japan, and into those taken by local government offices, these latter being the main institutions concerned with industrial cluster policy. Also surveyed by this study were the nationwide local offices of the Japan External Trade Organization (JETRO), which provides revitalzation support to locally-based small industries and enterprises.

METI, local government offices, and JETRO trade information centers were chosen as the targets of the study for the following reasons. As the main source of policy support for the industrial cluster projects now underway in Japan, METI has a good grasp of regional policy contexts and regional peculiarities. Local government offices, which cooperate closely with METI on the implementation of industrial cluster plans, are important policy-related institutions, and have a good grasp of the history and present circumstances of economic activity within the areas for which they are responsible. JETRO operates 36 trade information centers throughout Japan, in addition to their Tokyo and Osaka headquarters; as an institution concerned with the actual implementation of policy measures, JETRO maintains regular face-to-face contact with those people who are able to articulate the needs of local enterprises. As an organization that develops projects for revitalizing local economies, JETRO has intimate knowledge of the enterprises that help to activate local economies; of the availability of information at local levels, and of the circumstances of local economies and locally-based small companies. Moreover JETRO, as an independent administrative entity under law, occupies a position of neutrality with respect to companies and public institutions. These characteristics allow JETRO to have a thorough understanding of the conditions within industrial clusters.

1.3. Objects of the survey

	Sent	Returned	Return ratio
Ministry of economics, technology, and industry branches	24[16]	23[17]	95.8%
Local governments	49	37	75.6%
JETRO local offices	36	30	83.3%

Method: mailing
Operated in August, 2005
Volume: four pages long, approximately 20 minutes to take a survey

Acknowledgments

The author wishes to acknowledge Dr. Jobaid Kabir of University of Texas at Austin of United States for his valuable advice to this research. Also, the author expresses his appreciation to Mr. Mitsuhisa Katsumoto of METI Headquarter and staff of METI Local offices, for their valuable advice and contributions to his questionnaire survey. Finally, I am grateful to Dr. Akifumi Kuchiki and Dr. Yasushi Ueki of IDE-JETRO for their extensive comments.

Notes

1. In addition to those listed in the main text, there are also Sonobe, Tetsushi, and Keijiro Otsuka. 2004. *"Sangyo Hatten no Tsuru to Senryaku"* (Tools and Strategies for Industrial Development.) Chisen Shokan. Also, Hamamatsu Shinkin and Shinkin Central Bank Research Institute, eds. 2004. *"Sangyou Kurasutaa to Chiiki Kasseika"* (Industrial Clusters and Regional Revitalization.) Doyukan.
2. See Industrial Cluster Research Group, ed. May 2005. *"Sangyou Curasutaa Kenkyuukai Hokokusho"* (Study Report on Industrial Clusters.) METI. p. 16.
3. See p. 2 of footnote 2 reference material.
4. Porter, M.E. 1998. *On Competition*. Harvard Business School Press. pp. 199–200.
5. On the other hand, concerning the difference between cluster theory and conventional industrial agglomeration theory, Masahisa Fujita, an advocate of spatial economics, states, "Industrial agglomeration theory clarifies the mechanisms by which influences form to cause industrial agglomeration. Industrial clusters clarify the mechanisms of innovation creation in industrial agglomerations." (Masahisa Fujita 2003, pp. 212–213). In short, the decisive point is whether or not innovation creation takes places within an industrial agglomeration. When a cluster is formed, intellectual value such as technology, know-how, and knowledge pass through a horizontal network and are distributed swiftly, while at the same time, brisk innovations are triggered by mechanisms of competition and collaboration, enabling mobile responses to changes in the business environment. Currently, among those implementing local industrial

policy, strategy based on "industrial cluster theory" is considered valid. The aims of cluster strategy are to optimize the industrial organization of the overall region and to establish the competitive predominance of the region where industrial agglomerations are located. These aims are closely connected to the profit earned by firms in the agglomeration, which shows that the aims developed from the phenomenon of agglomeration development in economic theory are premised on cluster development. See Yamazaki, Akira. 2005. "*Sangyou Kurasutaa no Igi to Gendaiteki Kadai*" (Significance of Industrial Clusters and Current Issues). *Soshiki Kagaku*. Soshiki Gakkai. p. 11. Also see Yamazaki, Akira. 2002. "*Kurautaa Senryaku*" (Cluster Strategy). Yuhikaku. p. 4.

6. In the figure, the statistics for each cluster are organized by bundling prefectural industrial statistics into the region to which the prefecture belongs. Statistics were used only from prefectures where industrial cluster plans are underway. The data for "industrial shipment amount" and "added value" from 2002, the year after the Industrial Cluster Plans were implemented, and 2004, the year of the latest data, were used.
7. Concerning the outline of projects and the content of business in each sector, see the "Industrial Cluster Plan" of each region.
8. R&D of advanced medical equipment has been conducted to meet the needs of the healthcare field, including ergonomic "prevention," "diagnosis," and "treatment," using optic technologies such as super bright LEDs (light-emitting diodes) developed independently by Yamaguchi University.
9. Such as the Hiroshima central bio-cluster creation project, "Yamaguchi-Ube Medical Innovation Cluster."
10. A matter of special note in the Kinki Bio-Related Industry Project is that bio-ventures have appeared which conduct R&D jointly with large companies. Anges MG, Inc., which is famous as a successful university-launched venture, developed basic research on gene therapy being conducted by the university under the leadership of Ryuichi Morishita of the medical research department in the graduate school of Osaka University. Anges MG was set up to conduct R&D for pharmaceuticals marketing. Professor Morishita considered venture companies to be a necessary presence to act as a bridge between the seeds of university research and large companies because, during the past 10 to 20 years, the gap between large Japanese companies and universities has been widening due to globalization, and because projects without definite prospects for success were being abandoned. Anges MG is developing treatments involving gene therapy drugs for regeneration of blood vessels, together with, and with development funding from, a large pharmaceutical company. After the drugs are marketed, the tie-up agreement stipulates that the pharmaceutical company will pay Anges MG a fixed royalty on sales. As a result of this tie-up, the type of business model is becoming known in Japan in which a university-launched venture company that creates drugs works jointly with a large company to develop those drugs. Kodama, Toshihiro. 2004. "*Tama, Kinki, Kyuushuu no Hikaku ni Miru Sangyou Kurasutaa Keikaku no Kadai*" (Issues in Industrial Cluster Plans Seen Through a Comparison of Tama, Kinki, and Kyushu). Column No. 0128. Research Institute of Economy Trade and Industry.
11. Eco-towns take measures to construct environmentally friendly cities by effectively utilizing the existing industrial infrastructure, etc.
12. For the prototype model and the theoretical background of the flowchart model, see Chapter 1, Section 2 of "*Sangyou Kurasutaa Seisaku no Furochaato Apuroochii*

Purototaipu Moderu no Teiji" (Presentation of Prototype Model for the Flowchart Approach to Industrial Cluster Policy).

13. JETRO, the Japan External Trade Organization, is a government-related organization that works to promote mutual trade and investment between Japan and the rest of the world. Originally established in 1958 to promote Japanese exports, JETRO's core focus in the twenty-first century has shifted toward promoting foreign direct investment into Japan and helping small- and medium-sized Japanese firms maximize their global export potential. JETRO has 73 overseas offices in 55 countries worldwide, as well as 38 offices in Japan, including their Tokyo and Osaka headquarters. See outline and network on the JETRO's website (http://www.jetro.go.jp/en/jetro/network/).
14. "Technopolis" was a policy implemented by MITI (now METI) in 1980. It is a "strategy for introducing the vitality of advanced technological industries to local cultures, traditions, and natural areas by creating towns that harmonize 'industry' (advanced technological industrial groups), 'academia' (academic research institutions and pilot research institutions), and 'residences' (pleasant, stimulating living environments), to simultaneously achieve the goals of boosting knowledge-intensive industries and increasing added value in the industrial structure (i.e., building of a technologically creative country) and of developing outlying areas for the twenty-first century (permanent residence plan)." This definition is invariably incorporated in the development plan of each region, and numerous efforts have been undertaken to realize it.
15. Japan's industrial location policy from the mid-1950s to the mid-1970s focused mainly on the location of industry (i.e., the means of production). However, starting in the mid-1980s, software and services rapidly grew in economic importance, and development of the so-called "knowledge industry" in corporate research institutes, information processing departments, the software industry, and the design industry became important in revitalizing local economies. However, because this "knowledge industry" was concentrated mainly in the Tokyo area, the government decided to relocate it to outlying areas to promote the balanced development of the country. Thus, in 1988, the "Law on Promotion of Agglomeration of Specified Business that Contribute to the Advancement of Local Industries" (commonly known as the Know-how Location Law) was established. In the 26 regions nationwide where agglomeration promotion plans based on the Know-how Location Law were approved, various businesses have been promoted for the advancement of industry.
16. To cover all projects being implemented by all Ministry of Economics, Technology, and Industry branches nationwide, questionnaires were distributed to the staff at the branches in charge of each project. For this reason, the number of questionnaires differs from the number of branches (15).
17. In this survey, replies were received from the staff of all of the projects except one.

Bibliography

Ansoff, H. Igor. *The New Corporate Strategy, Revised Edition*. New York: Wiley, 1988.
Chesbrough, H. *Open Innovation*. Boston: Harvard Business School Press, 2003.
Christtensen, C. M. *Seeing What's Next*. Boston: Harvard Business School Press, 2004.

Fujita, M. "Kukan Keizaigaku No Shiten Kara Mita Sangyo Kurasuta Seisaku No Igi To Kadai." Ishikura, Y. et al. (eds) *Nihon No Sangyo Kurasuta Senryaku*. Yuhikaku Tokyo, 2003.

Fujita, M. and J. F. Thisse. *Economics of Agglomeration: Cities, Industrial Location and Regional Growth*. Cambridge: Cambridge University Press, 2002.

Hamamatsu Shinyo Kinko et al., *Sangyo Kurasuta To Chiiki Kasseika*. Doyukan Tokyo, 2004.

Ishikura, Y. et al., *Nihon No Sangyo Kurasuta Senryaku*. Yuhikaku Tokyo, 2003.

Itami, K, Shigeru, M and Takeo, K. *Sangyo Syuseki No Honshitsu*. Yuhikaku Tokyo, 1998.

Itoh, H. 1976. "Weber Kogyoricchiron Nyumon" Daimeidoh Press Tokyo.

Kuchiki, A. *Agglomeration of Exporting Firms in Industrial Zones in Northern Vietnam: Industrial Agglomeration*. Facts and Lessons for Developing Counties, Institute of Developing Economies JETRO, 2003.

Kuchiki A. "Flowchart Approach to Asia's Industrial Cluster Policy" *Industrial Cluster In Asia: Analyses of Their Competition and Cooperation. eds.* Kuchiki Akifumi and Masatsugu Tsuji. Institute of Developing Economies JETRO, 2004.

Kuchiki A. "Theoretical Models Based on a Flowchart Approach to Industrial Cluster Policy" *Discussion Paper No.33*. Institute of Developing Economies JETRO, 2005A.

Kuchiki A. "Theory of a Flowchart Approach to Industrial Cluster Policy", *Discussion Paper No.36*. Institute of Developing Economies JETRO, 2005B.

Local Branches, Ministry of Economy, Trade and Industry METI. *Sangyo Kurasuta Keikaku*. Ministry of Economy, Trade and Industry (METI) 2001.

Marshall, A. *Principles of Economics*. London: Macmillan, 1890.

Ministry of Economy, Trade and Industry METI. *Shin Sangyo Sozo Senryaku*. Ministry of Economy, Trade and Industry METI, 2004.

Nation Institute of Science and Technology Policy. *Kagaku Gijutu Hakusyo*. National Institute of Science and Technology Policy, 1998.

Nihon Chusyo Kigyo Gakkai Ronsyu. *Asia Shinjidai No Chusyo Kigyou*. Dohyukan, 2004.

Piore, M. J. and. Sabel, C.F. *The Second Industrial Divide*. New York: Basic Books, 1984.

Porter, M. E. *On Competition*. Boston: Harvard Business School Press, 1998.

Sangyo Kurasuta Research Group. *Sangyo Kurasuta Kenkyukai Hokokusyo*. Ministry of Economy, Trade and Industry METI, 2005.

Saxenian A. *Regional Advantage: Culture and Competition in Silicon Valley and Route 128*. Cambridge, MA: Harvard University Press, 1994.

Sonobe, T. and K. Otsuka. *Sangyo Hatten No Tsuru To Senryaku*. Chisensyokan, 2004.

Tsusyosangyo Research Group. *Tsusyosangyo Sesakushi*. Ministry of Economy, Trade and Industry METI, 1994.

Weber, A. *Kogyo Ricchi Ron*. Trans. Shinohara Taizo. Daimeidoh Press Tokyo, 1986.

Yamazaki, A. *Kurasuta Senryaku*. Yuhikaku Collection, 2002.

Yamazaki, A.. "Sangyo Kurasuta No Igi To Gendaiteki Kadai." *Soshiki Kagaku Soshiki Gakkai*, 2005: Vol. 38 No. 3 pp. 4–14.

Yoshida, K. and M. Nakanishi. "Factors Underlying the Formation of Industrial Clusters in Japan and Industrial Cluster Policy: A Quantitative Survey." *Discussion Paper Series* 2005: no.45, IDE-JETRO.

7

An Empirical Examination of the Flowchart Approach to Industrial Clustering: Case Study of Greater Bangkok, Thailand

Masatsugu Tsuji, Shoichi Miyahara, and Yasushi Ueki

1. Introduction

Industrial agglomeration, or clustering, is currently a hot topic in economics. There are two reasons for this. The first reason stems from the development of economic theory. There is a growing literature of the "new economic geography" initiated by Krugman (1995), Fujita, Krugman and Venables (1999), and Fujita and Thisse (2002). These new fields have expanded the boundaries of economic theory to new frontiers. The second reason relates to practical applications, such as economic policy. With the development of information and communication technology, commodities, human resources, and funds can move anywhere in the world in a relatively short time; the tides of so-called globalization are shaking all economies, developed and developing. This trend has resulted in the common practice of trying to attract foreign or even domestic resources located in different regions into an economy by stimulating a region's economic development. In order to invite foreign firms, for instance, policy makers must offer benefits to these firms; in doing so, they must understand the nature of agglomeration. Successful examples of such policies can be found in East Asian economies, one current example of the ongoing "East Asian Miracle."

In our previous researches, such as Kagami and Tsuji (2007) and Kuchiki and Tsuji (2005), we broadly analyze industrial agglomeration or clustering in Asia and the Americas. One positive result of these books has been to formulate a process of industrial agglomeration in developing economies, which is referred to as the "flowchart approach." This approach can be used to explain agglomeration in East Asia, and will be useful in crafting policy measures aimed at encouraging agglomeration.

Any theory must be able to verify its applicability to the actual economy. The objective of this paper is to examine the verifiability of the flowchart approach with an empirical study. In so doing, we selected the region of Thailand referred to as Greater Bangkok, and conducted mail surveys and in depth interviews to collect data for quantitative analysis. Based on the data obtained, we analyze the nature and characteristics of agglomeration in the region, and make an attempt to verify the flowchart approach.

The structure of the chapter is as follows: in Section 2, we explain the mail survey in Bangkok and show the nature of Bangkok's agglomeration. In Section 3, we provide a summary of the analysis. Mail surveys are not a rigorous method of analysis, and cannot obtain precise results. Thus, we apply heuristic estimations such as ordered probit models. The method and results of estimation will be presented in Section 4, and in Section 5 we examine verifiability of the flowchart approach by comparing the results obtained in Section 4. The final section provides brief conclusions and directions for future research.

2. Results of mail survey

2.1. Mail survey

The objective of this mail survey was to obtain and analyze data in order to verify the effectiveness of the flowchart approach to industrial cluster policy. We formulated a questionnaire based on Kuchiki (2005). Prior to the survey, in October 2005, we conducted a pre-test as well as face-to-face interviews with several Thai, Japanese, and other firms located around Greater Bangkok. In the interviews, we asked managers about the historical background of their Bangkok operations, the amount of their firms' investments in Bangkok, their reasons for establishing their Bangkok offices, the benefits and difficulties of doing businesses in Bangkok, future business plans, and desired government policies, etcetera. Information thus obtained was useful for improving the draft of the questionnaires. The 8-page questionnaires are shown in the Appendix.

Then questionnaires were sent in October, 2005 to 1,600 companies (1,000 Japanese, 300 other foreign and 300 native Thai companies) and we received 143 valid responses (117 from Japanese firms, 9 from other foreign firms, and 17 from Thai companies). The average response rate was 8.9% (11.7% for Japanese, 3.0% for other foreign, and 5.7% for Thai companies) (Table 7.1).[1]

2.2. Characteristics of the respondent firms

Here we summarize information obtained from our mail survey, which is the basis of the rigorous econometric analysis presented later in the chapter.

Table 7.1 Number of questionnaires dispatched and number of valid respondents to the questionnaire

	Dispatch	Valid response	Response rate
Japanese	1,000	117	11.7%
Thai	300	17	5.7%
Others	300	9	3.0%
Total	1,600	143	8.9%

Source: authors

First, profiles of the respondents are briefly summarized. In the next subsection, factors important in the firms' choice of business location are reviewed.

2.2.1. Year of business foundation

More than half of the firms surveyed started their Bangkok operations during or after the latter half of the 1980s. The rate of business establishment was greatest during the period of 1986–1990, followed by the years 1991–1995. By country of origin, most Thai companies were established between the latter half of the 1960s and the first half of the 1980s; a majority of Japanese and other foreign firms came to Thailand after the mid-1980s (Table 7.2).

2.2.2. Generation of the present CEO

More than 40% of the chief executive officers (CEOs) of the respondent companies are fourth or later generation, while 23.8% of the founders are still active in their companies. Foreign firms, including Japanese multinationals, are on their fourth or later CEOs. On the other hand, more than 40% of the CEOs of Thai local firms are the founders of those firms, even though their firms were generally established earlier in Bangkok than multinational companies (Table 7.3).

2.2.3. Size of firms: amount of capital

Surveyed firms are divided into two general groups by amount of capital. More than half have 400 million yen or less, while 20.3% are capitalized at over 2 billion yen. By national origin, about half of the Thai CEOs run heavier capitalized firms with more than 2 billion yen of capital (Table 7.4). This implies that many SMEs (small and medium-sized enterprises) are also established in Greater Bangkok; this is relevant to our examination of the flowchart approach, in which the relationship among large firms and SMEs plays a crucial role.

Table 7.2 Number of respondents by year of business foundation

(Total Respondents = 100)

	Total		**Japanese**		**Thai**		**Others**	
	No.	**% of total**	**No.**	**% of total**	**No.**	**% of total**	**No.**	**% of total**
Before 1950	10	7.0	7	6.0	2	11.8	1	11.1
1951~1955	6	4.2	6	5.1	0	0.0	0	0.0
1956~1960	3	2.1	3	2.6	0	0.0	0	0.0
1961~1965	3	2.1	2	1.7	0	0.0	1	11.1
1966~1970	7	4.9	4	3.4	3	17.6	0	0.0
1971~1975	8	5.6	5	4.3	2	11.8	1	11.1
1976~1980	3	2.1	1	0.9	2	11.8	0	0.0
1981~1985	5	3.5	3	2.6	2	11.8	0	0.0
1986~1990	36	25.2	33	28.2	1	5.9	2	22.2
1991~1995	24	16.8	21	17.9	1	5.9	2	22.2
1996~2000	22	15.4	17	14.5	3	17.6	2	22.2
2001~2005	12	8.4	12	10.3	0	0.0	0	0.0
N.A.	4	2.8	3	2.6	1	5.9	0	0.0
Total	143	100.0	117	100.0	17	100.0	9	100.0

Source: author.

Table 7.3 Number of respondents by generation of the present CEO (Total for Each Generation = 100)

	Total		Japanese		Thai		Others	
	No.	% of total	No.	% of total	No.	% of total	No.	% of total
Founder	34	23.8	27	23.1	7	41.2	0	0.0
2nd CEO	16	11.2	15	12.8	0	0.0	1	11.1
3rd CEO	19	13.3	18	15.4	1	5.9	0	0.0
4th or later	61	42.7	51	43.6	5	29.4	5	55.6
N.A.	13	9.1	6	5.1	4	23.5	3	33.3
Total	143	100.0	117	100.0	17	100.0	9	100.0

Source: author.

Table 7.4 Number of respondents by capital amount (10,000 yen)

	Total		Japanese		Thai		Others	
	No.	% of total	No.	% of total	No.	% of total	No.	% of total
10,000 or less	47	32.9	41	35.0	4	23.5	2	22.2
10,001~20,000	6	4.2	6	5.1	0	0.0	0	0.0
20,001~30,000	15	10.5	13	11.1	2	11.8	0	0.0
30,001~40,000	11	7.7	9	7.7	2	11.8	0	0.0
40,001~50,000	7	4.9	7	6.0	0	0.0	0	0.0
50,001~60,000	6	4.2	5	4.3	1	5.9	0	0.0
60,001~70,000	1	0.7	1	0.9	0	0.0	0	0.0
70,001~80,000	1	0.7	1	0.9	0	0.0	0	0.0
80,001~90,000	3	2.1	2	1.7	0	0.0	1	11.1
90,001~100,000	1	0.7	1	0.9	0	0.0	0	0.0
100,001~110,000	1	0.7	1	0.9	0	0.0	0	0.0
120,001~130,000	1	0.7	1	0.9	0	0.0	0	0.0

Continued

Table 7.4 Continued

	Total		Japanese		Thai		Others	
	No.	**% of total**	**No.**	**% of total**	**No.**	**% of total**	**No.**	**% of total**
130,001~140,000	1	0.7	1	0.9	0	0.0	0	0.0
150,001~160,000	1	0.7	1	0.9	0	0.0	0	0.0
160,001~170,000	2	1.4	1	0.9	0	0.0	1	11.1
170,001~180,000	1	0.7	0	0.0	0	0.0	1	11.1
190,001~200,000	1	0.7	1	0.9	0	0.0	0	0.0
200,001 and over	29	20.3	19	16.2	8	47.1	2	22.2
N.A.	8	5.6	6	5.1	0	0.0	2	22.2
Total	143	100.0	117	100.0	17	100.0	9	100.0

Source: author.

2.2.4. Size of firms: sales volume in 2004

Sales volume is another index of firm size. The distribution of sales volume among respondents is similar to that of capital amount. More Thai companies belong to the smaller ranges of sales volume, in comparison with capital amount (Table 7.5).

2.2.5. Size of firms: total number of employees

A similar distribution pattern among the respondent firms can be found in their total number of employees. More than half of respondents have fewer than 300 employees. As is the case for capital amount, there are a few large Thai enterprises in the group of top employers (Table 7.6).

2.2.6. Business fields

More than 70% of respondents are involved in the manufacturing sector. Some 67.5% of the Japanese companies, 77.8% of the other foreign companies, and all of the Thai firms have some sort of manufacturing-related businesses (note: respondents were asked to choose one or more business fields). The wholesale industry is the second most represented. Services are important businesses fields for the Japanese firms; many Japanese firms have already entered into Bangkok markets closely related to business support services and construction (Table 7.7). This seems to be consistent with the flowchart approach, since not only those related firms that supply parts and

Table 7.5 Number of respondents by sales for the year 2004 (10,000 yen)

	Total		Japanese		Thai		Others	
	No.	% of total	No.	% of total	No.	% of total	No.	% of total
100,000 or less	44	30.8	38	32.5	6	35.3	0	0.0
100,001~200,000	16	11.2	15	12.8	1	5.9	0	0.0
200,001~300,000	11	7.7	10	8.5	0	0.0	1	11.1
300,001~400,000	6	4.2	3	2.6	2	11.8	1	11.1
400,001~500,000	4	2.8	3	2.6	0	0.0	1	11.1
500,001~600,000	2	1.4	1	0.9	1	5.9	0	0.0
600,001~700,000	3	2.1	3	2.6	0	0.0	0	0.0
700,001~800,000	5	3.5	3	2.6	1	5.9	1	11.1
800,001~900,000	4	2.8	3	2.6	0	0.0	1	11.1
900,001~1000,000	2	1.4	1	0.9	1	5.9	0	0.0
1,000,001~1,100,000	2	1.4	2	1.7	0	0.0	0	0.0
1,200,001~1,300,000	2	1.4	2	1.7	0	0.0	0	0.0
1,300,001~1,400,000	3	2.1	3	2.6	0	0.0	0	0.0
1,400,001~1,500,000	2	1.4	2	1.7	0	0.0	0	0.0
1,500,001~1,600,000	2	1.4	2	1.7	0	0.0	0	0.0
1,600,001~1,700,000	1	0.7	1	0.9	0	0.0	0	0.0
1,700,001~1,800,000	1	0.7	1	0.9	0	0.0	0	0.0
1,900,001~2,000,000	1	0.7	1	0.9	0	0.0	0	0.0
2,000,001 and over	19	13.3	13	11.1	3	17.6	3	33.3
N.A.	13	9.1	10	8.5	2	11.8	1	11.1
Total	143	100.0	117	100.0	17	100.0	9	100.0

Source: author.

materials but also those that provide business support services come to a region after the establishment of large firms there, or after a certain level of agglomeration of firms has been reached.[2]

Among the firms in the manufacturing sector, about 25% and 16.5% deal with automobile-related and electronic-related products, respectively.[3]

Table 7.6 Number of respondents by number of total employees

	Total		Japanese		Thai		Others	
	No.	% of total	No.	% of total	No.	% of total	No.	% of total
1~100	33	23.1	30	25.6	2	11.8	1	11.1
101~200	18	12.6	13	11.1	4	23.5	1	11.1
201~300	20	14.0	18	15.4	0	0.0	2	22.2
301~400	10	7.0	8	6.8	1	5.9	1	11.1
401~500	6	4.2	4	3.4	1	5.9	1	11.1
501~600	2	1.4	2	1.7	0	0.0	0	0.0
601~700	3	2.1	2	1.7	1	5.9	0	0.0
701~800	4	2.8	3	2.6	1	5.9	0	0.0
801~900	4	2.8	2	1.7	2	11.8	0	0.0
901~1,000	3	2.1	3	2.6	0	0.0	0	0.0
1,001~1,100	1	0.7	1	0.9	0	0.0	0	0.0
1,101~1,200	4	2.8	3	2.6	1	5.9	0	0.0
1,301~1,400	1	0.7	1	0.9	0	0.0	0	0.0
1,701~1,800	1	0.7	1	0.9	0	0.0	0	0.0
1,801~1,900	1	0.7	1	0.9	0	0.0	0	0.0
2,001 and over	20	14.0	14	12.0	4	23.5	2	22.2
N.A.	12	8.4	11	9.4	0	0.0	1	11.1
Total	143	100.0	117	100.0	17	100.0	9	100.0

Source: author.

Following these are chemicals (14.6%), food (9.7%), metals (9.7%), steel (8.7%) and machinery and tools (8.7%). These results seem to largely reflect the fact that the majority of respondents were Japanese firms. Higher percentages of the Thai firms are engaged in steel (23.5%), food (17.6%) and textiles (17.6%) (Table 7.8).

Table 7.7 Number of respondents by business field

	Total		Japanese		Thai		Others	
	No.	% of total	No.	% of total	No.	% of total	No.	% of total
Manufacturing	103	72.0	79	67.5	17	100.0	7	77.8
Wholesale	14	9.8	9	7.7	3	17.6	2	22.2
Retail	7	4.9	6	5.1	1	5.9	0	0.0
Transportation	6	4.2	6	5.1	0	0.0	0	0.0
Construction	8	5.6	8	6.8	0	0.0	0	0.0
Finance, insurance	5	3.5	5	4.3	0	0.0	0	0.0
Corporate services	6	4.2	5	4.3	1	5.9	0	0.0
Individual services	3	2.1	2	1.7	1	5.9	0	0.0
Information technology	1	0.7	1	0.9	0	0.0	0	0.0
Telecommunications	3	2.1	3	2.6	0	0.0	0	0.0
Others	7	4.9	3	2.6	1	5.9	3	33.3
N.A.	2	1.4	2	1.7	0	0.0	0	0.0
Total	143	100.0	117	100.0	17	100.0	9	100.0

Source: author.

Table 7.8 Number of respondents by Manufactured Product

	Total		Japanese		Thai		Others	
	No.	% of total	No.	% of total	No.	% of total	No.	% of total
Food	10	9.7	6	7.6	3	17.6	1	14.3
Textiles	7	6.8	4	5.1	3	17.6	0	0.0
Wood	1	1.0	1	1.3	0	0.0	0	0.0
Paper	3	2.9	0	0.0	2	11.8	1	14.3
Synthetic resin, rubber	6	5.8	6	7.6	0	0.0	0	0.0

Continued

Table 7.8 Continued

	Total		Japanese		Thai		Others	
	No.	% of total	No.	% of total	No.	% of total	No.	% of total
Ceramic, rock, and Sand	2	1.9	2	2.5	0	0.0	0	0.0
Steel	9	8.7	5	6.3	4	23.5	0	0.0
Non-ferrous metals	3	2.9	3	3.8	0	0.0	0	0.0
Metals	10	9.7	9	11.4	1	5.9	0	0.0
Machinery and tools	9	8.7	9	11.4	0	0.0	0	0.0
Computers, computer parts	4	3.9	4	5.1	0	0.0	0	0.0
Other electronics	17	16.5	16	20.3	0	0.0	1	14.3
Cars, auto parts	25	24.3	23	29.1	2	11.8	0	0.0
Other transportation-related machinery	7	6.8	7	8.9	0	0.0	0	0.0
Chemicals, plastics	15	14.6	11	13.9	2	11.8	2	28.6
Other	20	19.4	7	8.9	10	58.8	3	42.9
N.A.	1	1.0	1	1.3	0	0.0	0	0.0
Total	103	100.0	79	100.0	17	100.0	7	100.0

Source: author.

3. Summary of the survey results

3.1. Establishment in greater Bangkok

The year respondents established their Bangkok offices is the most important variable in this analysis. Question 5 asks when a company established its first Bangkok office or facility. More than half of respondents began operating in Bangkok during or after the latter half of the 1980s. The number of establishments was greatest within the period of 1986–1990, followed by the years 1991–1995 and 1996–2000 (Table 7.9). This implies that agglomeration in the Bangkok area was not a smooth process, but has experienced peaks and troughs. In the later part of this paper, we analyze the nature and characteristics of agglomeration in this area in greater detail by independently examining several specific time periods.

About 50.4% of the Bangkok offices surveyed are subsidiaries of Japanese and other transnational companies. Some 30.8% and 7.0% of these are classified as headquarters and branches, respectively. A majority of these establishments carry out activities associated with sales (74.8%), accounting (69.9%), human resources (64.3%), marketing (59.4%), purchasing (56.6%) and production (52.4%), whereas fewer offices in Bangkok are responsible for research and development (R&D) (23.1%) and logistics (38.5%) (Tables 7.10 and 7.11).

Table 7.9 Year of establishment of Bangkok operations

	Total		Japanese		Thai		Others	
	No.	% of total	No.	% of total	No.	% of total	No.	% of total
Before 1950	4	2.8	1	0.9	2	11.8	1	11.1
1951~1955	1	0.7	1	0.9	0	0.0	0	0.0
1956~1960	2	1.4	2	1.7	0	0.0	0	0.0
1961~1965	2	1.4	1	0.9	0	0.0	1	11.1
1966~1970	8	5.6	5	4.3	3	17.6	0	0.0
1971~1975	8	5.6	5	4.3	2	11.8	1	11.1
1976~1980	5	3.5	1	0.9	4	23.5	0	0.0
1981~1985	7	4.9	5	4.3	2	11.8	0	0.0
1986~1990	38	26.6	36	30.8	0	0.0	2	22.2
1991~1995	25	17.5	21	17.9	2	11.8	2	22.2
1996~2000	25	17.5	21	17.9	2	11.8	2	22.2
2001~2005	11	7.7	11	9.4	0	0.0	0	0.0
Total	143	100.0	117	100.0	17	100.0	9	100.0

Source: author.

Table 7.10 Corporate form of the Bangkok operations

	Total		Japanese		Thai		Others	
	No.	**% of total**	**No.**	**% of total**	**No.**	**% of total**	**No.**	**% of total**
Headquarters	44	30.8	23	19.7	17	100.0	4	44.4
Branch	10	7.0	8	6.8	0	0.0	2	22.2
Subsidiary	85	59.4	82	70.1	0	0.0	3	33.3
N.A.	4	2.8	4	3.4	0	0.0	0	0.0
Total	143	100.0	117	100.0	17	100.0	9	100.0

Source: author.

Table 7.11 Function(s) carried out in Bangkok

	Total		Japanese		Thai		Others	
	No.	**% of total**	**No.**	**% of total**	**No.**	**% of total**	**No.**	**% of total**
Sales	107	74.8	84	71.8	15	88.2	8	88.9
Marketing	85	59.4	62	53.0	15	88.2	8	88.9
R&D	33	23.1	23	19.7	7	41.2	3	33.3
Production	75	52.4	63	53.8	10	58.8	2	22.2
Purchasing	81	56.6	61	52.1	14	82.4	6	66.7
Logistics	55	38.5	40	34.2	11	64.7	4	44.4
Accounting	100	69.9	78	66.7	14	82.4	8	88.9
Human affairs	92	64.3	73	62.4	13	76.5	6	66.7
General affairs	93	65.0	73	62.4	12	70.6	8	88.9
Others	11	7.7	7	6.0	3	17.6	1	11.1
N.A.	5	3.5	5	4.3	0	0.0	0	0.0
Total	143	100.0	117	100.0	17	100.0	9	100.0

Source: author.

Both market and supplier countries of the Bangkok operations (Question 4) are concentrated in a few countries. Their main market countries are China, Indonesia and Vietnam. These top three market countries account for 76% of respondents' sales. On the other hand, the main supplier countries are Japan and Thailand. 74% of the surveyed offices' purchases originate in these two countries.

3.2. Reasons for establishing operations in Bangkok

Question 8 asks about the reasons behind companies' decision to locate in Bangkok at the time Bangkok operations were initiated. Markets located near Bangkok were the most important factor affecting the decision to locate in Bangkok. This factor is followed by the presence of transportation infrastructure (roads and ports, etc.), other physical infrastructure (electricity, offices, etc.), tax and investment promotion policies, availability of professionals, social environment quality (culture, crime rates, etc.), institutional factors such as soundness of the financial system, and the legal system. Less important were access to cutting-edge technologies, physical environment quality (climate, pollution, etc.), and founders' personal choice (born in Bangkok, etc.). Higher priority was given to infrastructure, policies, and access to markets and inputs (Table 7.12). These observations from the mail survey provide important information in selecting the explanatory variables in our econometric estimation.

3.3. Current importance of and satisfaction with Bangkok operations

Question 9 asks firms how important Bangkok is for them as a business location. Some 86.0% of respondents recognize their Bangkok operations as important: 67.8% of them named it as very important and 18.2% as somewhat important. This awareness is common especially in the sectors of finance and insurance, corporate services, telecommunications, and information technology.

As for Question 10, which asks how satisfied firms are with Bangkok as a business location, 28.7% replied that they are very satisfied with current Bangkok operations, while 44.1% were somewhat satisfied, making a total 72.8% of the respondents satisfied with their operations in Bangkok.

3.4. Current importance of and satisfaction with Bangkok conditions: details

We also asked more detailed questions regarding the importance of, and firm's satisfaction with, Bangkok business conditions in Question 10 (How important are the following factors in your company's decision to continue/expand its Bangkok operations? And how satisfied are you with the current condition of each of these factors in Bangkok?). The respondents identified transportation infrastructure, IT infrastructure, markets located near Bangkok,

Table 7.12 Importance of the factors as the reasons for establishing operations in Bangkok (number and percentage of "important" + "somewhat important"answer)

	Total (N=143)		Japanese (N=117)		Thai (N=17)		Others (N=9)	
	No.	% of total	No.	% of Japanese	No.	% of Thai	No.	% of others
(1) Favorable government policies regarding taxes and investment	85	59.4	71	60.7	7	41.2	7	77.8
(2) Favorable government policies regarding trade	70	49.0	57	48.7	6	35.3	7	77.8
(3) Transportation infrastructure (roads, ports, etc.)	94	65.7	72	61.5	15	88.2	7	77.8
(4) Other physical infrastructure (electricity, offices, etc)	92	64.3	70	59.8	13	76.5	9	100.0
(5) IT infrastructure (broadband, etc.)	57	39.9	40	34.2	10	58.8	7	77.8
(6) Better financial system	70	49.0	53	45.3	10	58.8	7	77.8
(7) Better legal system	69	48.3	54	46.2	8	47.1	7	77.8
(8) Better protection of intellectual property rights	40	28.0	28	23.9	5	29.4	7	77.8
(9) Market located near Bangkok	97	67.8	80	68.4	12	70.6	5	55.6
(10) Advantage for export	61	42.7	47	40.2	10	58.8	4	44.4
(11) Suppliers/ subcontractors located near Bangkok	76	53.1	61	52.1	11	64.7	4	44.4
(12) Request by large company/related company or govt.	54	37.8	45	38.5	6	35.3	3	33.3
(13) Unskilled labor market	63	44.1	57	48.7	2	11.8	4	44.4
(14) Skilled labor market	71	49.7	52	44.4	12	70.6	7	77.8
(15) Availability of professionals	83	58.0	61	52.1	13	76.5	9	100.0

Continued

Table 7.12 Continued

	Total (N=143)		Japanese (N=117)		Thai (N=17)		Others (N=9)	
	No.	% of total	No.	% of Japanese	No.	% of Thai	No.	% of others
(16) Other companies from the same country are located here	44	30.8	35	29.9	7	41.2	2	22.2
(17) Access to cutting-edge information / technology (spillovers)	27	18.9	17	14.5	8	47.1	2	22.2
(18) Environmental quality (climate, pollution, etc.)	28	19.6	19	16.2	3	17.6	6	66.7
(19) Social quality (culture, crime rates, etc.)	78	54.5	62	53.0	9	52.9	7	77.8
(20) Living environment (shopping, entertainment, etc)	59	41.3	44	37.6	9	52.9	6	66.7
(21) Rents, land prices, cost of living	64	44.8	49	41.9	9	52.9	6	66.7
(22) Congestion	42	29.4	33	28.2	5	29.4	4	44.4
(23) Founder's personal choice (born in Bangkok, etc.)	29	20.3	18	15.4	10	58.8	1	11.1

Source: author.

other physical infrastructure such as electricity and offices, and favorable government policies regarding taxes and investment, as important factors. This indicates that better business conditions related to infrastructure and policies are required (Table 7.13).

Some respondents related their needs for world-class legal and tax systems, easy access to reliable and up-to-date information, and a sound governance system with transparency and accountability. They also revealed their demands for management systems suitable to the Thai business environment, including management of local employees and employee-employer relationships, and rigorous cost management capabilities appropriate for major overseas production bases.

Table 7.13 Current importance of factors in respondents' decision to continue/expand their Bangkok operations (number and percentage of "important" + "somewhat important"answers)

	Total (N=143)		Japanese (N=117)		Thai (N=17)		Others (N=9)	
	No.	% of total	No.	% of Japanese	No.	% of Thai	No.	% of others
(1) Favorable government policies regarding taxes and investment	92	64.3	75	64.1	10	58.8	7	77.8
(2) Favorable government policies regarding trade	82	57.3	66	56.4	10	58.8	6	66.7
(3) Transportation infrastructure (roads, ports, etc.)	109	76.2	86	73.5	15	88.2	8	88.9
(4) Other physical infrastructure (electricity, offices, etc.)	95	66.4	72	61.5	15	88.2	8	88.9
(5) IT infrastructure (broadband, etc.)	93	65.0	68	58.1	16	94.1	9	100.0
(6) Better financial system	86	60.1	66	56.4	14	82.4	6	66.7
(7) Better legal system	89	62.2	71	60.7	11	64.7	7	77.8
(8) Better protection of intellectual property rights	52	36.4	39	33.3	8	47.1	5	55.6
(9) Market located near Bangkok	91	63.6	74	63.2	12	70.6	5	55.6
(10) Advantage for export	71	49.7	52	44.4	13	76.5	6	66.7
(11) Suppliers/ subcontractors located near Bangkok	86	60.1	69	59.0	13	76.5	4	44.4
(12) Request by large company/related company or govt.	35	24.5	26	22.2	6	35.3	3	33.3
(13) Unskilled labor market	72	50.3	62	53.0	6	35.3	4	44.4
(14) Skilled labor market	89	62.2	66	56.4	15	88.2	8	88.9
(15) Availability of professionals	98	68.5	73	62.4	16	94.1	9	100.0

Continued

Table 7.13 Continued

	Total (N=143)		Japanese (N=117)		Thai (N=17)		Others (N=9)	
	No.	% of total	No.	% of Japanese	No.	% of Thai	No.	% of others
(16) Other companies from the same country are located here	46	32.2	38	32.5	6	35.3	2	22.2
(17) Access to cutting-edge information/ technology (spillovers)	27	18.9	13	11.1	11	64.7	3	33.3
(18) Environmental quality (climate, pollution, etc.)	50	35.0	34	29.1	10	58.8	6	66.7
(19) Social quality (culture, crime rates, etc.)	87	60.8	69	59.0	12	70.6	6	66.7
(20) Living environment (shopping, entertainment, etc.)	67	46.9	50	42.7	13	76.5	4	44.4
(21) Rents, land prices, cost of living	77	53.8	60	51.3	12	70.6	5	55.6
(22) Congestion	68	47.6	53	45.3	10	58.8	5	55.6
(23) Founder's personal choice (born in Bangkok, etc.)	27	18.9	15	12.8	11	64.7	1	11.1

Source: author.

Regarding satisfaction with the various factors, most respondents are satisfied with the living environment (shopping, entertainment, cost of living, etc.), markets located near Bangkok, social environment quality (Thai society, culture, national character), advantages for export, suppliers/subcontractors located near Bangkok, and so on. In contrast, they expressed their dissatisfaction with traffic congestion, poor physical environment quality, weak protection of intellectual property, and an incomplete legal system. In summary, respondents expressed satisfaction with access to market and suppliers, but dissatisfaction with infrastructure and institutions (Table 7.14).

In general, respondent companies are content with the quality of labor, but discontent with the Thai government and institutional factors. Regarding labor quality, the diligence of Thai people receives high praise, although their level of training and education are rather unsatisfactory. Problems

Table 7.14 Current satisfaction with factors in respondents' decision to continue/expand their Bangkok operations (number and percentage of "satisfied" + "somewhat satisfied" answers)

	Total (N=143)		Japanese (N=117)		Thai (N=17)		Others (N=9)	
	No.	% of total	No.	% of Japanese	No.	% of Thai	No.	% of others
(1) Favorable government policies regarding taxes and investment	48	33.6	37	31.6	5	29.4	6	66.7
(2) Favorable government policies regarding trade	82	57.3	66	56.4	10	58.8	6	66.7
(3) Transportation infrastructure (roads, ports, etc.)	109	76.2	86	73.5	15	88.2	8	88.9
(4) Other physical infrastructure (electricity, offices, etc)	95	66.4	72	61.5	15	88.2	8	88.9
(5) IT infrastructure (broadband, etc.)	93	65.0	68	58.1	16	94.1	9	100.0
(6) Better financial system	86	60.1	66	56.4	14	82.4	6	66.7
(7) Better legal system	89	62.2	71	60.7	11	64.7	7	77.8
(8) Better protection of intellectual property rights	52	36.4	39	33.3	8	47.1	5	55.6
(9) Market located near Bangkok	91	63.6	74	63.2	12	70.6	5	55.6
(10) Advantage for export	71	49.7	52	44.4	13	76.5	6	66.7
(11) Suppliers/ subcontractors located near Bangkok	86	60.1	69	59.0	13	76.5	4	44.4
(12) Request by large company/related company or govt.	35	24.5	26	22.2	6	35.3	3	33.3
(13) Unskilled labor market	72	50.3	62	53.0	6	35.3	4	44.4
(14) Skilled labor market	89	62.2	66	56.4	15	88.2	8	88.9
(15) Availability of professionals	98	68.5	73	62.4	16	94.1	9	100.0
(16) Other companies from the same country are located here	46	32.2	38	32.5	6	35.3	2	22.2

Continued

Table 7.14 Continued

	Total (N=143)		Japanese (N=117)		Thai (N=17)		Others (N=9)	
	No.	% of total	No.	% of Japanese	No.	% of Thai	No.	% of others
(17) Access to cutting-edge information / technology (spillovers)	27	18.9	13	11.1	11	64.7	3	33.3
(18) Environmental quality (climate, pollution, etc)	50	35.0	34	29.1	10	58.8	6	66.7
(19) Social quality (culture, crime rates, etc.)	87	60.8	69	59.0	12	70.6	6	66.7
(20) Living environment (shopping, entertainment, etc.)	67	46.9	50	42.7	13	76.5	4	44.4
(21) Rents, land prices, cost of living	77	53.8	60	51.3	12	70.6	5	55.6
(22) Congestion	68	47.6	53	45.3	10	58.8	5	55.6
(23) Founder's personal choice (born in Bangkok, etc.)	27	18.9	15	12.8	11	64.7	1	11.1

Source: author.

caused by poor government capabilities and institutional limitations are regarded as time-consuming, as are uncertain custom clearance procedures, ill-planned road construction that causes heavy traffic jams, and so on. There was also a comment on the existence of unfair treatment against Japanese firms by government organizations.

3.5. Current problems with the Bangkok business environment

Question 13 asks about problems firms have doing business in Bangkok. Some 32.9% of the companies surveyed are aware of some sort of problems in their Bangkok businesses. More precisely, 14.0% have experienced major difficulties, and 18.9% have experienced slight difficulties (Table 7.15). Replies according to industry are shown in Table 7.16.

Human resource management is a serious issue for companies in Bangkok.[4] Human resource-related problems include difficulties in communication among employees because of language barriers, difficulties in recruiting skilled labor, the difficulty of headhunting for excellent human resources,

Table 7.15 Existence of problems regarding businesses in Bangkok

	Total	Yes	A little	Not sure	Not much	Not at all	N.A.
Total	143	20	27	33	40	9	14
	100.0%	14.0%	18.9%	23.1%	28.0%	6.3%	9.8%
Japanese	117	19	18	32	32	3	13
	100.0%	16.2%	15.4%	27.4%	27.4%	2.6%	11.1%
Thai	17	1	7	1	5	2	1
	100.0%	5.9%	41.2%	5.9%	29.4%	11.8%	5.9%
Others	9	0	2	0	3	4	0
	100.0%	0.0%	22.2%	0.0%	33.3%	44.4%	0.0%

Source: author.

Table 7.16 Existence of problems regarding businesses in Bangkok

	Total	Yes	A little	Not sure	Not much	Not at all	N.A.
Total	143	20	27	33	40	9	14
	100.0%	14.0%	18.9%	23.1%	28.0%	6.3%	9.8%
Manufacturing	103	16	18	19	33	6	11
	100.0%	15.5%	17.5%	18.4%	32.0%	5.8%	10.7%
Wholesale	14	3	2	4	4	0	1
	100.0%	21.4%	14.3%	28.6%	28.6%	0.0%	7.1%
Retail	7	1	2	1	1	1	1
	100.0%	14.3%	28.6%	14.3%	14.3%	14.3%	14.3%
Transportation	6	0	4	2	0	0	0
	100.0%	0.0%	66.7%	33.3%	0.0%	0.0%	0.0%
Construction	8	0	0	5	2	0	1
	100.0%	0.0%	0.0%	62.5%	25.0%	0.0%	12.5%
Finance, insurance	5	1	1	1	1	0	1
	100.0%	20.0%	20.0%	20.0%	20.0%	0.0%	20.0%
Corporate services	6	1	0	2	0	1	2
	100.0%	16.7%	0.0%	33.3%	0.0%	16.7%	33.3%
Individual services	3	0	0	2	0	1	0
	100.0%	0.0%	0.0%	66.7%	0.0%	33.3%	0.0%

Continued

Table 7.16 Continued

	Total	Yes	A little	Not sure	Not much	Not at all	N.A.
Information technology	1	0	0	0	1	0	0
	100.0%	0.0%	0.0%	0.0%	100.0%	0.0%	0.0%
Telecommunications	3	0	1	1	1	0	0
	100.0%	0.0%	33.3%	33.3%	33.3%	0.0%	0.0%
Others	7	0	0	2	0	5	0
	100.0%	0.0%	0.0%	28.6%	0.0%	71.4%	0.0%
N.A.	2	0	1	0	1	0	0
	100.0%	0.0%	50.0%	0.0%	50.0%	0.0%	0.0%

Source: author.

and the higher turnover rate of unskilled labor. These issues cause increases in wage rates and in costs such as employee training and protection of secret information.

Many other complaints were given about the legal system and about taxation, especially tax- and customs-related rules and their operations. Although amendments of laws such as tax code and labor law, for example, have been announced throughout Thailand's governments, these amendments are unclear and force companies to refer to the responsible government sections again and again. Regarding custom clearance procedures, there is a lack of consistent and uniform interpretation and operation of the code. Custom officers administrate their duties in an arbitrary manner, so that importers are sometimes forced to pay unexpected high tariffs.

3.6. Future plans

Some 81.1% of respondents are certain that they will continue their business operations in Bangkok, and 9.8% are "probably" sure. In total, more than 90% of the surveyed companies stated their intention to continue operations in Bangkok (Table 7.17).

Among the companies that expressed their intention to continue their Bangkok operations, 35.4% intend very strongly to increase their investments and 19.2% probably intend this; more than half are willing to increase their investments in Bangkok (Table 7.18). Three firms, however, replied "No" to Question 14 (Will your company continue its business in Bangkok?), stating that they will not continue business in Bangkok because of "supplier

Table 7.17 Future prospects for continuing businesses in Bangkok

	Total	Yes	A little	Not sure	Not much	Not at all	N.A.
Total	143	116	14	6	3	0	4
	100.0%	81.1%	9.8%	4.2%	2.1%	0.0%	2.8%
Manufacturing	103	80	10	6	3	0	4
	100.0%	77.7%	9.7%	5.8%	2.9%	0.0%	3.9%
Wholesale	14	12	1	1	0	0	0
	100.0%	85.7%	7.1%	7.1%	0.0%	0.0%	0.0%
Retail	7	7	0	0	0	0	0
	100.0%	100.0%	0.0%	0.0%	0.0%	0.0%	0.0%
Transportation	6	4	2	0	0	0	0
	100.0%	66.7%	33.3%	0.0%	0.0%	0.0%	0.0%
Construction	8	7	1	0	0	0	0
	100.0%	87.5%	12.5%	0.0%	0.0%	0.0%	0.0%
Finance, insurance	5	5	0	0	0	0	0
	100.0%	100.0%	0.0%	0.0%	0.0%	0.0%	0.0%
Corporate services	6	6	0	0	0	0	0
	100.0%	100.0%	0.0%	0.0%	0.0%	0.0%	0.0%
Individual services	3	3	0	0	0	0	0
	100.0%	100.0%	0.0%	0.0%	0.0%	0.0%	0.0%
Information technology	1	1	0	0	0	0	0
	100.0%	100.0%	0.0%	0.0%	0.0%	0.0%	0.0%
Telecommunications	3	3	0	0	0	0	0
	100.0%	100.0%	0.0%	0.0%	0.0%	0.0%	0.0%
Others	7	7	0	0	0	0	0
	100.0%	100.0%	0.0%	0.0%	0.0%	0.0%	0.0%
N.A.	2	1	1	0	0	0	0
	100.0%	50.0%	50.0%	0.0%	0.0%	0.0%	0.0%

Source: author.

Table 7.18 Intention to increase investments in Bangkok (respondents = "yes" and "probably" answers to the question in Table 7.18)

	Total	Very much	Somewhat	Not sure	Not much	Not at all	N.A.
Total	130	46	25	40	9	6	4
	100.0%	35.4%	19.2%	30.8%	6.9%	4.6%	3.1%
Manufacturing	90	38	21	20	6	2	3
	100.0%	42.2%	23.3%	22.2%	6.7%	2.2%	3.3%
Wholesale	13	2	2	7	1	0	1
	100.0%	15.4%	15.4%	53.8%	7.7%	0.0%	7.7%
Retail	7	2	0	4	0	1	0
	100.0%	28.6%	0.0%	57.1%	0.0%	14.3%	0.0%
Transportation	6	1	1	3	1	0	0
	100.0%	16.7%	16.7%	50.0%	16.7%	0.0%	0.0%
Construction	8	1	2	3	0	2	0
	100.0%	12.5%	25.0%	37.5%	0.0%	25.0%	0.0%
Finance, insurance	5	3	0	2	0	0	0
	100.0%	60.0%	0.0%	40.0%	0.0%	0.0%	0.0%
Corporate services	6	1	1	2	0	1	1
	100.0%	16.7%	16.7%	33.3%	0.0%	16.7%	16.7%
Individual services	3	1	0	2	0	0	0
	100.0%	33.3%	0.0%	66.7%	0.0%	0.0%	0.0%
Information technology	1	0	0	1	0	0	0
	100.0%	0.0%	0.0%	100.0%	0.0%	0.0%	0.0%
Telecommunications	3	0	0	2	0	1	0
	100.0%	0.0%	0.0%	66.7%	0.0%	33.3%	0.0%
Others	7	2	2	1	2	0	0
	100.0%	28.6%	28.6%	14.3%	28.6%	0.0%	0.0%
N.A.	2	0	0	2	0	0	0
	100.0%	0.0%	0.0%	100.0%	0.0%	0.0%	0.0%

Source: author.

changes," "deterioration of the environment in Bangkok (including the social environment)," and "existence of better location(s) than Bangkok." Some 14.0% of the surveyed firms have plans to start new operations somewhere in Thailand (Table 7.19). The candidate sites include Rayong, Nakhon Ratchasima, Prachin Buri, Ayuthaya, Saraburi, Khon Kaen/Kanchanaburi, and others. About 21.0% plan to start new operations outside Thailand (Table 7.20). Prospected destinations of their investment outside Thailand include China, India, Vietnam, and other countries.

3.7. Requests for government policies

Question 17 asks about Thai government policies that help firms do business in Bangkok. The respondents' top prioritized areas are as follows: deregulation in various fields (36.4%), preferential tax measures for companies (34.3%), technical training of workers (27.3%), and decreasing traffic jams (26.6%). The three areas given top priority were deregulation (12.1%), preferential tax measures for companies (11.4%), and decreasing traffic jams (8.9%). Strong hopes for countermeasures against traffic congestion may be unique to Bangkok (Table 7.21 and 7.22).

Table 7.19 Reasons for no future prospects for continuing businesses in Bangkok

	Total		Japanese		Thai	
	No.	% of total	No.	% of total	No.	% of total
Head office's order	0	0.0%	0	0.0%	0	0.0%
Change in government regulation	0	0.0%	0	0.0%	0	0.0%
Market change	1	33.3%	1	100.0%	0	0.0%
Supplier change	1	33.3%	0	0.0%	1	50.0%
Major customer change	0	0.0%	0	0.0%	0	0.0%
Language problem	0	0.0%	0	0.0%	0	0.0%
Environment (including social environment)	1	33.3%	0	0.0%	1	50.0%
Rent and land	0	0.0%	0	0.0%	0	0.0%
Living costs	0	0.0%	0	0.0%	0	0.0%
Better location has been found	1	33.3%	0	0.0%	1	50.0%
Other	0	0.0%	0	0.0%	0	0.0%
Total	3	100.0%	1	100.0%	2	100.0%

Source: author.

Table 7.20 Existence of plans to start new operations in countries other than in Thailand

	Total	Yes	Not sure	No	N.A.
Total	143	30	35	66	12
	100.0%	21.0%	24.5%	46.2%	8.4%
Manufacturing	103	22	26	46	9
	100.0%	21.4%	25.2%	44.7%	8.7%
Wholesale	14	3	5	5	1
	100.0%	21.4%	35.7%	35.7%	7.1%
Retail	7	3	2	1	1
	100.0%	42.9%	28.6%	14.3%	14.3%
Transportation	6	3	0	3	0
	100.0%	50.0%	0.0%	50.0%	0.0%
Construction	8	1	2	4	1
	100.0%	12.5%	25.0%	50.0%	12.5%
Finance, insurance	5	0	1	3	1
	100.0%	0.0%	20.0%	60.0%	20.0%
Corporate services	6	3	0	3	0
	100.0%	50.0%	0.0%	50.0%	0.0%
Individual services	3	1	0	2	0
	100.0%	33.3%	0.0%	66.7%	0.0%
Information technology	1	0	0	1	0
	100.0%	0.0%	0.0%	100.0%	0.0%
Telecommunications	3	0	0	2	1
	100.0%	0.0%	0.0%	66.7%	33.3%
Others	7	1	3	3	0
	100.0%	14.3%	42.9%	42.9%	0.0%
N.A.	2	0	0	1	1
	100.0%	0.0%	0.0%	50.0%	50.0%

Source: author.

Table 7.21 The most important policies requested (total of the three most important)

	Total		Japanese		Thai		Others	
	No.	% of total	No.	% of total	No.	% of total	No.	% of total
Total	143	100.0%	117	100.0%	17	100.0%	9	100.0%
Tax cuts/ tax incentives	49	34.3%	43	36.8%	5	29.4%	1	11.1%
Deregulation	52	36.4%	50	42.7%	1	5.9%	1	11.1%
Lower interest rates	10	7.0%	9	7.7%	0	0.0%	1	11.1%
Lower lease prices	0	0.0%	0	0.0%	0	0.0%	0	0.0%
Subsidies	12	8.4%	10	8.5%	2	11.8%	0	0.0%
Improvements in education	39	27.3%	38	32.5%	1	5.9%	0	0.0%
R&D improvement	8	5.6%	8	6.8%	0	0.0%	0	0.0%
IT improvement	20	14.0%	19	16.2%	1	5.9%	0	0.0%
Office environment improvement	3	2.1%	3	2.6%	0	0.0%	0	0.0%
Congestion relief	38	26.6%	35	29.9%	3	17.6%	0	0.0%
Others	7	4.9%	7	6.0%	0	0.0%	0	0.0%
N.A.	58	40.6%	38	32.5%	12	70.6%	8	88.9%

Source: author.

Table 7.22 The most important policies requested

	Total		Japanese		Thai		Others	
	No.	% of total	No.	% of total	No.	% of total	No.	% of total
Total	143	100.0%	117	100.0%	17	100.0%	9	100.0%
Tax cuts/ tax incentives	30	21.0%	26	22.2%	3	17.6%	1	11.1%
Deregulation	19	13.3%	19	16.2%	0	0.0%	0	0.0%
Lower interest rates	1	0.7%	1	0.9%	0	0.0%	0	0.0%
Lower lease prices	0	0.0%	0	0.0%	0	0.0%	0	0.0%
Subsidies	2	1.4%	2	1.7%	0	0.0%	0	0.0%

Continued

Table 7.22 Continued

	Total		Japanese		Thai		Others	
	No.	% of total	No.	% of total	No.	% of total	No.	% of total
Improvements in education	10	7.0%	10	8.5%	0	0.0%	0	0.0%
R&D improvement	3	2.1%	3	2.6%	0	0.0%	0	0.0%
IT improvement	3	2.1%	3	2.6%	0	0.0%	0	0.0%
Office environment improvement	1	0.7%	1	0.9%	0	0.0%	0	0.0%
Congestion relief	9	6.3%	7	6.0%	2	11.8%	0	0.0%
Others	7	4.9%	7	6.0%	0	0.0%	0	0.0%
N.A.	58	40.6%	38	32.5%	12	70.6%	8	88.9%

Source: author.

4. Summary of implications of the mail survey results

4.1. Past agglomeration factors

According to the results of our mail survey, we can summarize factors of agglomeration in Greater Bangkok as follows: Markets located near Bangkok seems to have been the most important factor attracting firms to establish their Bangkok offices during the 1980s. This was followed by transportation infrastructure, other physical infrastructure, preferential treatments concerning taxes and investment, availability of professionals, social environment quality, and financial and legal systems. In contrast, less important are intra-industrial exchanges, physical environment quality, and founder's personal choice. Accordingly, important factors are found in infrastructure, policies, market conditions, and suppliers/subcontractor location. These factors are consistent with those suggested by the flowchart approach.

In 2005, 85.5% of respondent companies recognized the importance of their Bangkok operations. This awareness is present especially in the services sector, including finance, insurance, corporate services, telecommunications, and information technology businesses. A careful examination of the responses reveals the most important factors to be transportation infrastructure, IT infrastructure, markets located near Bangkok, other physical infrastructure, and favorable government policies regarding taxes and investment. The most important factors were those related to infrastructure, followed by policy factors.

When companies reply to questions related to investments in Bangkok, they pay more attention to markets located near Bangkok, infrastructure such as transportation, and other physical and social capital. IT infrastructure is also increasing in importance.

The respondents express a high degree of satisfaction with Bangkok as their business location; 72.8% of them gave positive replies. Among the factors affecting companies' decisions to continue/expand their Bangkok operations, they were satisfied with living environment, markets located near Bangkok, social environment quality, advantages for export, and procurement of parts and materials around Bangkok. They are discontented, however, with traffic congestion, deterioration of physical environment quality, insufficient protection of intellectual property rights, and an inadequate legal system. These "dissatisfied factors" are closely related to infrastructure and institutions, while those with which respondents are satisfied are associated with market conditions and suppliers/subcontractors located near Bangkok.

At the same time, 32.9% of the surveyed companies have experienced problems doing business in Bangkok. The problems pointed out are involved mainly with human resources: language barriers, securing of qualified human resources, head-hunting of skilled and professional workers, education programs and personnel evaluation systems, high turnover rate of employees, inadequate employee skills. The second group of serious problems is related to the complicated tax system and various ambiguities in legal interpretation.

Nevertheless, more than 90% of the companies intend to continue operations in Bangkok. Among the companies that express their intent to continue their Bangkok operations, more than half intend to increase investment in Bangkok.

The respondents want the Thai government to take the following actions: deregulation in various fields (36.4%), preferential tax measures for companies (34.3%), technical training of workers (27.3%), and decreasing traffic jams (26.6%). Strong hopes for countermeasures against traffic congestion may be a unique characteristic of Bangkok firms.

5. Estimation of agglomeration factors in greater Bangkok

5.1. Selection of variables for estimation

In the previous sections we described the nature of industrial agglomeration in Greater Bangkok. This section presents empirical estimations to verify the flowchart approach as a theory of industrial clustering. Firms consider many economic as well as non-economic factors of the particular areas where they establish their business activities. The flowchart approach, developed by Kuchiki (2005), (2007), summarizes the following factors causing agglomeration as follows: (i) domestic demand; (ii) export; and (iii) capacity building. Factor (iii) includes infrastructure, institutional factors such as legislation, deregulation, and other economic systems, human resources, and social factors such as living conditions, traditional religions and morals, etc. In order to empirically verify the flowchart approach, we construct the following hypothesis: When firms decide on the locations of their business operations, they take into account all factors related to the possible locations, including those suggested by the flowchart approach. The result of their decision-making process is revealed after they actually establish business operations in a certain location. In our survey, the results of this decision-making process are analyzed with regard to the year firms' Bangkok offices were established. Year of establishment, therefore, is our dependent variable, and we can examine relationships among year of establishment and the other explanatory variables indicated by the flowchart approach. Our main hypothesis is that the year of establishment of Bangkok offices is inversely related to firm size, which interprets the flowchart approach to predict that anchor firms come earlier, then related and affiliated firms follow. We also examine how factors attracting firms to Greater Bangkok changed according to different time periods. Firms' decision-making processes regarding establishing Bangkok offices were surely different before and after the East Asian Economic Crisis in 1997.

5.2. Determination of dependent variable

As explained earlier, the year of establishment of the Bangkok office is used as our dependent variable. The number of offices established in Greater Bangkok in different years is denoted in Table 7.23. According to this table, the number increased after 1987, when the Japanese economy enjoyed the economic boom called the "bubble economy," greatly increasing the costs of labor and land. In addition, the Japanese yen appreciated a huge amount in a short time. These conditions encouraged Japanese firms to establish their factories and offices in developing countries. Since 1987, the number of Bangkok offices increased greatly; in the 11 years from 1987 to 1997, 64 offices were established, which is about 57% of the total.

Thus, we intuitively divide our time domain into two regions, but we must verify the validity of this. Tsuji, Miyahara and Ishikawa (1999) utilized the stepwise-Chow test in order to identify the exact year of industrial transformation in the Japanese machine tool industry. Since this paper deals with cross-sectional data, we cannot use the Chow test. In order to identify the year of structural shift in the Greater Bangkok region's agglomeration, we use the following method. First, we fix a set of independent variables, which

Table 7.23 Year of establishment in Bangkok

Year of business foundation in Bangkok	Number of establishments	Percent	Cumulative percent
1882	1	0.89	0.89
1928	1	0.89	1.79
1932	1	0.89	2.68
1952	1	0.89	3.57
1964	1	0.89	4.46
1965	1	0.89	5.36
1966	2	1.79	7.14
1967	1	0.89	8.04
1968	1	0.89	8.93
1969	1	0.89	9.82
1970	3	2.68	12.5
1972	2	1.79	14.29
1973	3	2.68	16.96
1975	1	0.89	17.86
1976	1	0.89	18.75
1977	1	0.89	19.64
1980	2	1.79	21.43
1981	1	0.89	22.32
1984	2	1.79	24.11
1985	4	3.57	27.68
1986	1	0.89	28.57
1987	8	7.14	35.71
1988	7	6.25	41.96
1989	8	7.14	49.11
1990	8	7.14	56.25
1991	5	4.46	60.71
1992	4	3.57	64.29
1993	2	1.79	66.07
1994	2	1.79	67.86
1995	6	5.36	73.21
1996	7	6.25	79.46
1997	7	6.25	85.71
1998	2	1.79	87.5
1999	5	4.46	91.96
2000	2	1.79	93.75

Continued

Table 7.23 Continued

Year of business foundation in Bangkok	Number of establishments	Percent	Cumulative percent
2001	3	2.68	96.43
2002	1	0.89	97.32
2003	1	0.89	98.21
2004	1	0.89	99.11
2005	1	0.89	100
Total	112	100	–

Source: author.

will be defined later. Second, we identify the year (j in Equation (1) below) in which structural change in agglomeration supposedly occurred, and construct dependant variables based on this year (j) by the following formula:

$$\text{Year of establishing Bangkok office} = \begin{cases} 0, \text{ if year} < 1971 + (j-1) \\ 1, \text{ if year} > 1971 + j \\ \qquad \text{where } j = 1, 2, \ldots, L, \ldots 27. \end{cases} \quad (1)$$

Letting j vary from 1 to 27, this process produces 27 variables for each j. For each year j, we conduct a probit analysis. The results of recursive estimation are shown in Table 7.24. From Table 7.24, it follows that the fit is the best in 1987 in terms of log likelihood as well as pseudo R2, and we can conclude that this indicates that before and after 1987, factors affecting establishment of Bangkok offices probably changed; that is, structural change occurred in 1987. In addition, looking at Table 7.24, we can see that during 1988 and 1990, this fit suddenly becomes worse, and after 1991 to 1997, log likelihood and pseudo R2 become rather stable, but in 1997, when the East Asian Economic crisis occurred, the fit again improves. According to these observations, since structural changes also occurred in these periods, we can divide the period after 1987 into a number of sub-periods. These sub-period results will be used for our actual estimation.

5.3. Selection of independent variables

We made a careful attempt to order our questionnaires in order for them to represent the core of the flowchart approach; thus, the main questions ask about factors suggested by the flowchart approach. Independent variables are summarized as follows:

Amount of capital represents the size of firms (Question 1).

Table 7.24 Indication of industrial agglomeration

Year of business foundation in Bangkok	Log likelihood	Pseudo R2
1972	–9.89241	0.7656
1973	–14.775	0.6783
1974	–14.775	0.6783
1975	–16.0779	0.6847
1976	–13.2625	0.7476
1977	–13.9687	0.7416
1978	–13.9687	0.7416
1979	–13.9687	0.7416
1980	–17.0088	0.6935
1981	–22.0177	0.6216
1982	–22.0177	0.6216
1983	–22.0177	0.6216
1984	–22.7199	0.6179
1985	–23.5929	0.6186
1986	–21.2086	0.679
1987	–19.6461	0.7068
1988	–31.8862	0.5632
1989	–37.6668	0.5056
1990	–42.8973	0.4473
1991	–41.7518	0.456
1992	–37.9911	0.4937
1993	–40.4226	0.4462
1994	–37.5408	0.4767
1995	–39.4075	0.4397
1996	–38.7232	0.405
1997	–37.2026	0.3458
1998	–25.791	0.3905

Source: author.

Production and manufacturing represent functions of Bangkok operations (Question 7).

Questions regarding 13 factors such as infrastructure, institutions, marker conditions, labor, and industrial structure, that influenced the establishment of Bangkok offices, are asked in Question 8.[5] According to the discussions in the mail survey responses, some factors such as liking conditions in Bangkok are omitted, because strong relationships are not found.

Dummy variables are attached to manufacturing and wholesale industry for the type of Bangkok operations.[6]

A summary of statistics regarding these variables is indicated in Table 7.25.

5.4. Hypotheses

The flowchart approach gives clues to the following hypotheses concerning relationships between dependent variables, explanatory variables, and industrial agglomeration. We postulate the hypotheses as follows:

5.4.1. Hypothesis regarding business scale indicators

The flowchart approach indicates that larger companies would establish offices earlier than smaller ones. The accumulation of business resources in a particular location, such as infrastructures, institutional systems, and human resources, which is thought to be triggered by foreign direct investment (FDI) by large firms, can encourage the generation of other new industries and FDIs by smaller companies and suppliers. In this model, capital amount should have a negative correlation with year of business foundation in Bangkok (positive for companies that started their business operations earlier). On the other hand, positive correlations (positive for companies that started their business operations later) should be expected for the indicators of "request by large company/related company," "other companies from the same country are located here," and "access to cutting-edge information/technology (spillovers)," which are items in Question 8 of our questionnaire.

5.4.2. Hypothesis regarding the relationship with the indicators of Bangkok office's functions

Considering the background of the hollowing out of production bases in Japan and the effects of this on industrial agglomeration in Asia, functions of the Bangkok offices are thought to be shifting: during the early stage of agglomeration, Bangkok attracted firms mainly as a production base. As agglomeration proceeds, activities in a location will shift from simple functions such as production to higher-level ones, including R&D activities. Thus, the variable related to the function of Bangkok offices for "production" should be negatively correlated with the year of business establishment in Bangkok, and positively correlated for "R&D activities."

Table 7.25 Summary statistics

Variables		Mean	Standard deviation	Minimum	Maximum
Q5 (Explained variable)	Year of business foundation in Bangkok	1986.295	16.223	1882	2005
	Year of business foundation in Bangkok (case A)	0.714	0.454	0	1
	Year of business foundation in Bangkok (case B)	2.036	0.782	1	3
	Year of business foundation in Bangkok (case C)	1.857	0.642	1	3
Q1 Business scale	Capital amount (log-transformed)	10.17	2.27	3	16
	Year of business foundation	25.098	24.154	1	125
	Year of business foundation (squared)	1208.134	2794.168	1	15625
Q7	R&D	0.232	0.424	0	1
(Function)	Production	0.554	0.499	0	1
Q8 (Government policies)	Favorable government policies regarding taxes and investment	0.17	0.377	0	1
	Favorable government policies regarding trade	0.205	0.406	0	1
(Social capitals)	Transportation infrastructure (roads, ports, etc.)	0.107	0.311	0	1
(Market environment)	Other physical infrastructure (electricity, offices, etc.)	0.098	0.299	0	1
(Employment conditions)	IT infrastructure (broadband, etc.)	0.188	0.392	0	1
(Industrial structure)	Market located near Bangkok	0.161	0.369	0	1
	Market located near Bangkok	0.268	0.445	0	1
	Suppliers/subcontractors located near Bangkok	0.196	0.399	0	1

Continued

Table 7.25 Continued

Variables		Mean	Standard deviation	Minimum	Maximum
	Request by large company/related company or govt.	0.304	0.462	0	1
	Unskilled labor market	0.313	0.466	0	1
	Skilled labor market	0.205	0.406	0	1
	Other companies from the same country are located here	0.321	0.469	0	1
	Access to cutting-edge information/technology (spillovers)	0.357	0.481	0	1
Q2	Manufacturing	0.741	0.44	0	1
(Business fields)	Wholesale	0.107	0.311	0	1

Notes:

1. Year of business foundation in Bangkok (case A) is a variable that is equal to 0 for 1986 and before and 1 for 1987 and later.
2. Year of business foundation in Bangkok (case B) is a variable that is 0 if the year is before 1986, 1 if the year is between 1987 and 1997, and 3 if the year is after 1998.
3. Year of business foundation in Bangkok (case C) is a variable that is 1 if the year is before 1986, 2 if the year is between 1987 and 1994, and 3 if the year is after 1995.
4. R&D and production in Q7 are dummy variables that are equal to 1 if they fall under these categories respectively.
5. Each item in Q8 is 1 if respondents chose "very important" or "somewhat important", otherwise 0.
6. The number of observations = 112.

Source: author.

5.4.3. Hypothesis regarding the relationship between the indicators of infrastructure (social capital) and human resources

The development of infrastructure and labor availability are important factors that affect multinational companies' decision-making processes regarding overseas locations. According to the flowchart approach, further development of social capital, improvement in labor quality, and business specialization progress along with increasing agglomeration, leads to even further agglomeration by attracting more firms. In this sense, the availabilities of skilled labor and professionals become more important factors over time. Thus, it is expected that "infrastructure (social capital)" and "availability of unskilled labor" have negative correlations with the year of business establishment in Bangkok, whereas "availability of skilled labor" positively correlates with the dependent variable.

5.5. Models to be estimated

As explained in the previous section, we divide years of business foundation in Bangkok into two periods, the period through 1986 and the period after 1987, by focusing on a structural change in the factors affecting development of business operations in Bangkok. In addition, in order to examine these structural changes in more detail, we split the latter period into two sub-periods, because the East Asian Economic Crisis occurred in 1997, and because two peaks of establishments of Bangkok offices are found in Table 7.23. By considering the characteristics of these periods and sub-periods, we formulate the following three models for estimation:

Model A: This case divides years of business foundation in Bangkok into two periods, through 1986 and after 1987. A binomial probit estimation will be utilized.

Model B: Taking into account the effects of the Asian financial crisis, business foundation years are grouped into the following three sub-periods: (i) through 1986 (Group 1); (ii) between 1987 and 1997 (Group 2); and (iii) after 1998 (Group 3). We use for our estimates both an ordered probit model that reflects orders between the groups, and an ordinary multinomial probit model.

Model C: In the period after 1986, when companies rushed into Bangkok, there are two sub-periods that show sharp increases in the number of establishments: these periods are from 1987 to 1990 and from 1995 to 1997. This implies the existence of "first movers" and "followers" of industrial agglomeration; the differences in foundation year are thought to be attributable to differences in characteristics such as firm size and business field. The business establishment years can thus be divided into three sub-periods: through 1986 (Group 1); from 1987 to 1997 (Group 2); and after 1998 (Group 3). As in Model B, both an ordered probit and an ordinary multinomial probit model are utilized in Model C.

5.6. Results of estimation

5.6.1. Model A: binomial probit model

The result of estimation by the binomial probit model is summarized in Table 7.26. Firms starting operations in Bangkok through 1986 are referred to as Group 1A, whereas Group 2A are those that opened after 1987. We take Group 1A as a benchmark; thus, if the marginal effects of variables in the table have positive signs, the variables have positive effects relative to Group 2A (This is the same as stating that they have negative effects relative to Group 1A).[8]

Among the explanatory variables that are significant at 10 percent or less, factors positively affecting Group 1A (the "first-movers" to Bangkok; this implies that the variables have negative sign) are:

1% significant level:
- *Production* (marginal effect: 0.263); functions of the Bangkok offices
- *Wholesale* (marginal effect: 0.998); dummy variables for business fields

5% significant level:
- *Amount of capital* (marginal effect is 0.005); the indicator of business size
- *Favorable government policies regarding taxes and investment* (marginal effect: 0.976);
- *Other companies from the same country are located in Bangkok* (marginal effect: 0.753); variable for industrial structure.

10% significant level:
- *R&D* (marginal effect: 0.064); functions of the Bangkok offices

On the other hand, positive factors for the "follower"-type Group 2 are:

10% significant level
- *Market located near Bangkok* (marginal effect: 0.025);
- *Skilled labor market* (marginal effect: 0.133);
- *Access to cutting-edge information/technology (spill-overs)* (marginal effect: 0.049); Spill-over effect.

In accordance with the flowchart approach, we can interpret these estimates as follows. In the industrial agglomeration in Greater Bangkok, until 1986 larger-sized firms entered into the area earlier than smaller ones, aiming for production and wholesale. They were also attracted by favorable government policies regarding taxes and investment, and the previously existing industrial agglomeration, at the time when they made their investment decisions. Their Bangkok offices also carried out functions of R&D. During the second stage, following the large firms, smaller-sized firms of various industrial sectors

Table 7.26 Binomial probit model results

Variables	Coefficient	t-statistic	Marginal effect
Constant	11.254	3.24***	—
Capital amount (log-transformed)	–0.347	–2.02**	–0.005
Year of business foundation	–0.208	–3.41***	–0.003
Year of business foundation (squared)	0.001	2.6***	0
R&D	–1.348	–1.82*	–0.064
Production	–4.268	–2.63***	–0.263
Favorable government policies regarding taxes and investment	–5.437	–1.98**	–0.976
Favorable government policies regarding trade	4.954	1.76*	0.064
Transportation infrastructure (roads, ports, etc.)	–3.505	–1.17	–0.72
Other physical infrastructure (electricity, offices, etc.)	0.747	0.54	0.006
IT infrastructure (broadband, etc.)	–1.024	–1.14	–0.041
Market located near Bangkok	3.598	1.71*	0.025
Market located near Bangkok	1.419	1.29	0.015
Suppliers/subcontractors located near Bangkok	–1.533	-0.97	–0.093
Request by large company/related company or govt.	1.265	1.05	0.015
Unskilled labor market	–1.458	–1.3	–0.061
Skilled labor market	6.957	1.92*	0.133
Other companies from the same country are located here	–4.752	–2.32**	–0.753
Access to cutting-edge information/ technology (spillovers)	2.487	1.74*	0.049
Business field (manufacturing)	1.298	0.99	0.055
Business field (wholesale)	–6.34	–2.9***	–0.998
Log likelihood	–19.646		

Note: ***, **, and * are significant at 1%, 5%, and 10%, respectively.

Source: author.

entered this area, attracted by markets developed by the first-movers, high-quality labor forces, and information related to technological know-how. Except for R&D, which is typically thought to occur as a result of clustering during the later period, these estimation results largely support the flowchart approach, although infrastructure and institutions are not found to be significant.

5.6.2. Model B: multinomial probit model

In this model, two new groups are created for the period after 1987. Group 2B includes those established after 1987 but before the East Asian Economic Crisis in 1997, and Group 3B includes those established after the 1997 incident. Group 1B is the same as Group 1A in Model A. In making estimations for this kind of model, the ordered probit model is theoretically more suitable. We conducted two estimations, an ordered and a multinomial probit model, by comparing these results. It is clearly demonstrated that the latter model can explain our results better than the former, in terms of log likelihood (Table 7.27, 7.28, 7.29, and 7.30). The reason for this is that the latter loses less information due to the difference in effects of the explanatory variables among groups.

The estimates for this model have similar patterns to those for Case A. It should be noted that there are almost no differences in the signs of marginal effects between Group 2B and Group 3B (Table 7.29 and 7.30). This result implies that the Asian financial crisis did not seriously affect decision-making factors regarding business establishment in Greater Bangkok.[9]

Let us discuss briefly the interpretation of these estimations in terms of the flowchart approach. Among the explanatory variables, those positively affecting Group 1B (the "first-movers" to Bangkok) are:

1% significant level:
- *Production* (marginal effect: 0.051); functions of the Bangkok offices
- *Wholesale* (marginal effect: 0.275); dummy variables for business fields

Following variables are also found to be significant, although their significant levels are low:
- *Amount of capital* (marginal effect is 0.005);
- *Favorable government policies regarding taxes and investment* (marginal effect: 0.990);
- *Other companies from the same country are located in Bangkok* (marginal effect: 0.889).

10% significant level
- *R&D* (marginal effect: 0.051).

5.6.3. Model C: multinomial probit model

Because Model B produced the same results as Model A, in order to analyze how variables affect agglomeration differently in the different periods, we present another model with a different definition of time periods. By observing the peaks of business establishment years, we can specify the year 1994 as a boundary for defining the "second group," whose Bangkok operations were begun after 1987. We define offices established during 1987 and 1994

Table 7.27 Result for multinomial probit model
(case B: ordered probit)

Variables	Coefficient	t-statistic
Capital amount (log-transformed)	–0.03	–0.47
Year of business foundation	–0.09	–4.88***
Year of business foundation (squared)	0.001	3.37***
R&D	–0.112	–0.34
Production	–0.712	–1.73*
Favorable government policies regarding taxes and investment	–0.783	–1.45
Favorable government policies regarding trade	0.679	1.28
Transportation infrastructure (roads, ports, etc.)	–0.742	–1.26
Other physical infrastructure (electricity, offices, etc.)	0.135	0.25
IT infrastructure (broadband, etc.)	–	–
Market located near Bangkok	0.474	1.2
Market located near Bangkok	0.624	1.43
Suppliers/subcontractors located near Bangkok	–0.703	–1.38
Request by large company/related company or govt.	–0.239	–0.66
Unskilled labor market	–0.173	–0.43
Skilled labor market	1.134	2.44**
Other companies from the same country are located here	–0.477	–1.24
Access to cutting-edge information/ technology (spillovers)	0.827	2.35**
Business field (Manufacturing)	–0.166	–0.43
Business field (Wholesale)	–1.336	–2.47**
1st threshed	–3.212	
2nd threshed	–0.612	
Log likelihood	–68.883	

Note: ***, **, and * are significant at 1%, 5%, and 10%, respectively.

Source: author.

Table 7.28 Marginal effects in multinomial probit model (case B: ordered probit)

Variables	**Mairgnal effect**		
	Group 1	**Group 2**	**Group 3**
Capital amount (log-transformed)	0.008	–0.005	–0.003
Year of business foundation	0.024	–0.015	–0.009
Year of business foundation (squared)	0	0	0
R&D	0.03	–0.02	–0.01
Production	0.18	–0.105	–0.075
Favorable government policies regarding taxes and investment	0.248	–0.198	–0.05
Favorable government policies regarding trade	–0.146	0.055	0.092
Transportation infrastructure (roads, ports, etc.)	0.24	–0.196	–0.044
Other physical infrastructure (electricity, offices, etc.)	–0.034	0.02	0.014
IT infrastructure (broadband, etc.)	–	–	–
Market located near Bangkok	–0.106	0.047	0.06
Market located near Bangkok	–0.143	0.065	0.077
Suppliers/subcontractors located near Bangkok	0.217	–0.169	–0.048
Request by large company/related company or govt.	0.065	–0.044	–0.021
Unskilled labor market	0.047	–0.031	–0.016
Skilled labor market	–0.213	0.025	0.188
Other companies from the same country are located here	0.135	–0.094	–0.04
Access to cutting-edge information/technology (spillovers)	–0.194	0.094	0.1
Business field (manufacturing)	0.042	–0.025	–0.017
Business field (wholesale)	0.465	–0.406	–0.059

Note: ***, **, and * are significant at 1%, 5%, and 10%, respectively.

Source: author.

as Group 2C, and those after 1995 as Group 3C. Again, observing the results of our ordered probit and multinomial probit models (Table 7.31, 7.32, 7.33 and 7.34), we find that the latter is more explanatory than the former, and we thus conduct the latter.

According to the result of this estimation, there are considerable differences in the signs and values of variables between marginal effects for

Table 7.29 Result for multinomial probit model
(case B: multinomial probit)

Variables	Group 2		Group 3	
	Coefficient	t-statistic	Coefficient	t-statistic
Constant	14.283	3.02***	14.166	2.89***
Capital amount (log-transformed)	–0.409	–1.75*	–0.423	–1.6
Year of business foundation	–0.261	–3.06***	–0.372	–3.86***
Year of business foundation (squared)	0.002	2.29**	0.002	3.27***
R&D	–1.957	–1.94*	–0.828	–0.68
Production	–5.841	–2.67***	–6.918	–3.01***
Favorable government policies regarding taxes and investment	–8.082	–1.79*	–7.48	–1.63
Favorable government policies regarding trade	7.63	1.67*	7.125	1.54
Transportation infrastructure (roads, ports, etc.)	–5.304	–1.34	–6.402	–1.55
Other physical infrastructure (electricity, offices, etc.)	0.117	0.07	0.606	0.32
IT infrastructure (broadband, etc.)	–	–	–	–
Market located near Bangkok	6.023	1.98**	4.249	1.32
Market located near Bangkok	1.539	0.98	2.677	1.5
Suppliers/subcontractors located near Bangkok	–2.16	–1.01	–4.222	–1.76*
Request by large company/ related company or govt.	1.406	0.88	1.44	0.84
Unskilled labor market	–2.447	–1.62	–2.471	–1.51
Skilled labor market	11.235	2.23**	11.761	2.3**
Other companies from the same country are located here	–7.688	–2.56**	–6.706	–2.17**
Access to cutting-edge information/technology (spillovers)	3.598	1.76*	3.843	1.78*
Business field (manufacturing)	1.755	0.99	2.382	1.26
Business field (wholesale)	–8.976	–2.9***	–9.99	–3.03***
Log likelihood	–48.378			

Notes:

1. ***, **, and * are significant at 1%, 5%, and 10%, respectively.
2. The items expressed by "–" were excluded from the explanatory variables because all of them become 0 in Group3.

Source: author.

Table 7.30 Marginal effects in multinomial probit model (case B: multinomial probit)

Variables	Marginal effect		
	Group 1	Group 2	Group 3
Capital amount (log-transformed)	0.004	0.002	–0.003
Year of business foundation	0.003	0.011	–0.014
Year of business foundation (squared)	0	0	0
R&D	0.051	–0.214	0.163
Production	0.275	–0.106	–0.17
Favorable government policies regarding taxes and investment	0.99	–0.902	–0.088
Favorable government policies regarding trade	–0.078	0.116	–0.038
Transportation infrastructure (roads, ports, etc.)	0.824	–0.709	–0.115
Other physical infrastructure (electricity, offices, etc)	–0.002	–0.07	0.072
IT infrastructure (broadband, etc.)	–	–	–
Market located near Bangkok	–0.029	0.152	–0.123
Market located near Bangkok	–0.013	–0.167	0.18
Suppliers/subcontractors located near Bangkok	0.116	0.04	–0.156
Request by large company/related company or govt.	–0.011	0.005	0.007
Unskilled labor market	0.093	–0.073	–0.02
Skilled labor market	–0.2	0.094	0.106
Other companies from the same country are located here	0.889	–0.836	–0.054
Access to cutting-edge information/ technology (spillovers)	–0.054	0.012	0.042
Business field (Manufacturing)	–0.059	–0.017	0.076
Business field (Wholesale)	0.999	–0.885	–0.114

Source: author.

Groups 2C and 3C. Significant explanatory variables are: production; wholesale; R&D; favorable government policies regarding trade; market located near Bangkok; skilled labor market; and access to cutting-edge information/ technology.

Again, as in the previous two models, production (activity) and wholesale (business field) have negative sign in the two periods after 1986, implying that firms that came earlier to Bangkok exhibited these characteristics.

Table 7.31 Result for multinomial probit model (case C: ordered probit)

Variables	Coefficient	t-statistic
Capital amount (log-transformed)	0.024	0.36
Year of business foundation	–0.089	–4.67***
Year of business foundation (squared)	0	2.96***
R&D	–0.4	–1.25
Production	–0.593	–1.4
Favorable government policies regarding taxes and investment	–0.885	–1.64
Favorable government policies regarding trade	0.896	1.69*
Transportation infrastructure (roads, ports, etc.)	–0.181	–0.28
Other physical infrastructure (electricity, offices, etc.)	–0.012	–0.02
IT infrastructure (broadband, etc.)	–0.958	–2.43**
Market located near Bangkok	0.141	0.37
Market located near Bangkok	0.161	0.37
Suppliers/subcontractors located near Bangkok	0.037	0.07
Request by large company/related company or govt.	–0.285	–0.79
Unskilled labor market	–0.17	–0.44
Skilled labor market	1.887	3.37***
Other companies from the same country are located here	–0.677	–1.73*
Access to cutting–edge information/technology (spillovers)	0.845	2.31**
Business field (manufacturing)	0.447	1.03
Business field (wholesale)	–1.781	–3.16***
1st threshed	–2.463	
2nd threshed	–0.647	
Log likelihood	–76.347	

Note: ***, **, and * are significant at 1%, 5%, and 10%, respectively.

Source: author.

Favorable government policies regarding taxes and investment and transportation infrastructure (roads and ports, etc.) also show negative signs in these periods, from which we deduce that first movers pay great attention to these. Other variables, such as (i) amount of capital, (ii) favorable government policies regarding trade, (iii) unskilled labor market; (iv) skilled labor market; (v) other companies from the same country located in Bangkok, and

Table 7.32 Marginal effects in multinomial probit model (case C: ordered probit)

Variables	**Marginal effect**		
	Group 1	**Group 2a**	**Group 2b**
Capital amount (log–transformed)	–0.006	–0.001	0.007
Year of business foundation	0.021	0.004	–0.026
Year of business foundation (squared)	0	0	0
R&D	0.107	–0.002	–0.105
Production	0.138	0.035	–0.174
Favorable government policies regarding taxes and investment	0.269	–0.076	–0.193
Favorable government policies regarding trade	–0.163	–0.138	0.301
Transportation infrastructure (roads, ports, etc.)	0.047	0.002	–0.049
Other physical infrastructure (electricity, offices, etc.)	0.003	0.001	–0.003
IT infrastructure (broadband, etc.)	0.292	–0.084	–0.208
Market located near Bangkok	–0.032	–0.01	0.042
Market located near Bangkok	–0.037	–0.01	0.048
Suppliers/subcontractors located near Bangkok	–0.009	–0.002	0.011
Request by large company/related company or govt.	0.073	0.006	–0.078
Unskilled labor market	0.042	0.005	–0.048
Skilled labor market	–0.262	–0.377	0.639
Other companies from the same country are located here	0.182	–0.007	–0.175
Access to cutting–edge information/ technology (spillovers)	–0.18	–0.082	0.262
Business field (manufacturing)	–0.119	0.002	0.117
Business field (wholesale)	0.605	–0.345	–0.26

Source: author.

(vi) access to cutting-edge information/technology, have negative signs during 1987–1994, but positive after 1995. This indicates that these variables became more important in the period after 1995. Regarding firm size, during 1987–1994 small-sized firms established offices in the region, while after 1995 larger firms tended to establish offices. Variables such as "market located near Bangkok" and "request by large company/related company or government" have positive signs during 1987–1995, while negative after

Table 7.33 Result for multinomial probit model
(case C: multinomial probit)

Variables	Group 2a		Group 2b	
	Coefficient	t statistic	Coefficient	t statistic
Constant	15.95	3.190***	16.076	3.07***
Capital amount (log–transformed)	–0.594	–2.45**	–0.502	–2**
Year of business foundation	–0.179	–1.89*	–0.354	–3.9***
Year of business foundation (squared)	0	–0.13	0.002	3.01***
R&D	–1.856	–1.77*	–2.391	–2.1**
Production	–6.972	–2.88***	–6.815	–2.84***
Favorable government policies regarding taxes and investment	–8.656	–2.41**	–8.988	–2.41**
Favorable government policies regarding trade	7.199	1.95*	7.951	2.1**
Transportation infrastructure (roads, ports, etc.)	–5.427	–1.51	–4.875	–1.34
Other physical infrastructure (electricity, offices, etc)	1.501	0.78	1.774	0.8
IT infrastructure (broadband, etc.)	–1.891	–1.45	–4.064	–2.54**
Market located near Bangkok	5.54	2**	4.142	1.46
Market located near Bangkok	2.56	1.68*	1.471	0.88
Suppliers/subcontractors located near Bangkok	–2.396	–1.13	–1.884	–0.83
Request by large company/ related company or govt.	3.07	1.81*	2.469	1.43
Unskilled labor market	–2.357	–1.48	–2.099	–1.31
Skilled labor market	8.962	2.12**	11.484	2.65***
Other companies from the same country are located here	–7.378	–2.76***	–6.815	–2.52**
Access to cutting–edge information/technology (spillovers)	3.679	1.97**	3.788	1.93*
Business field (manufacturing)	1.888	1.01	3.149	1.58
Business field (wholesale)	–9.399	–2.99***	–12.573	–3.57***
Log likelihood	–53.179			

Note: ***, **, and * are significant at 1%, 5%, and 10%, respectively.

Source: author.

Table 7.34 Marginal effects in multinomial probit model (case C: multinomial probit)

Variables	**Marginal effect**		
	Group 1	**Group 2a**	**Group 2b**
Capital amount (log–transformed)	0.009	–0.03	0.022
Year of business foundation	0.004	0.047	–0.051
Year of business foundation (squared)	0	–0.001	0.001
R&D	0.114	0.078	–0.193
Production	0.482	–0.292	–0.19
Favorable government policies regarding taxes and investment	0.999	–0.521	–0.478
Favorable government policies regarding trade	–0.109	–0.149	0.258
Transportation infrastructure (roads, ports, etc.)	0.863	–0.505	–0.358
Other physical infrastructure (electricity, offices, etc.)	–0.01	–0.072	0.082
IT infrastructure (broadband, etc.)	0.197	0.321	–0.518
Market located near Bangkok	–0.033	0.364	–0.331
Market located near Bangkok	–0.021	0.299	–0.278
Suppliers/subcontractors located near Bangkok	0.132	–0.202	0.07
Request by large company/related company or govt.	–0.037	0.184	–0.147
Unskilled labor market	0.105	–0.124	0.019
Skilled labor market	–0.201	–0.459	0.659
Other companies from the same country are located here	0.899	–0.548	–0.351
Access to cutting–edge information/ technology (spillovers)	–0.08	0.012	0.068
Business field (manufacturing)	–0.139	–0.232	0.371
Business field (wholesale)	1	–0.442	–0.558

Source: author.

1995. This indicates that during 1987–1995, firms tend to open offices in Bangkok for those two reasons.

This result for Model C confirms the possibility of structural change in some of the determining factors of industrial agglomeration formation in Thailand after 1994. Based on the estimation results of our three models, the characteristics and structure of the industrial agglomeration in Greater Bangkok are summarized in Figure 7.1.

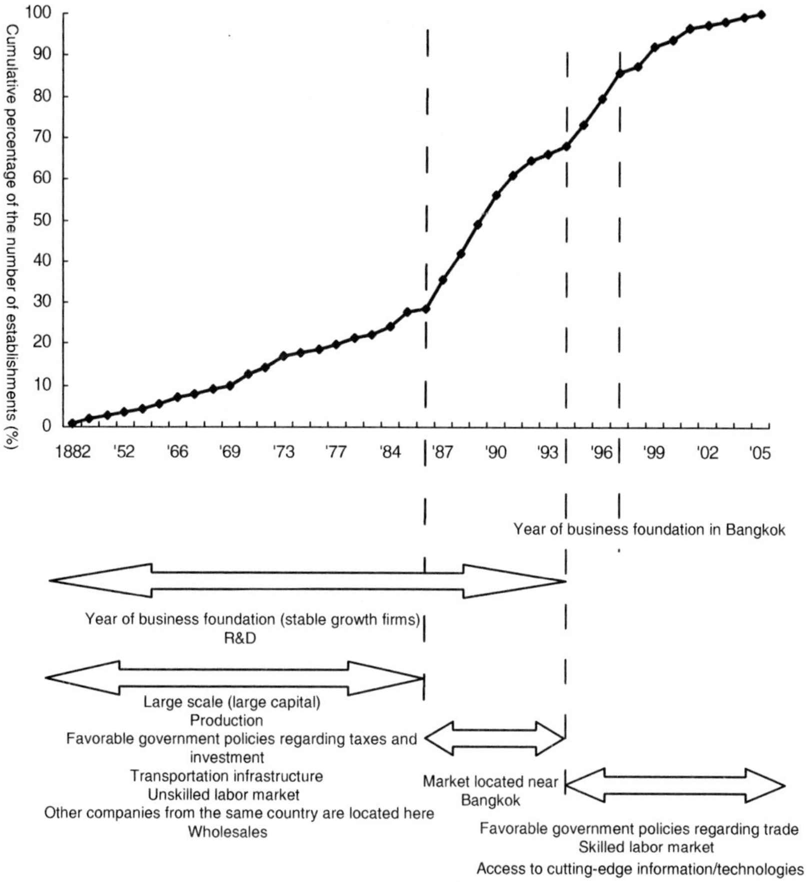

Figure 7.1 Structure of the industrial agglomeration in the greater Bangkok, Thailand

Notes

1: Factors with positive coefficients in Table 7.35 were drawn (Marginal effects are negative during the periods without arrows).

2: The dot-line drawn from the year 1997 does not intersect with any arrows. This means that there was no change in the determinants of establishments of offices in Bangkok.

Source: Prepared by the author.

Table 7.35 Summary of estimates of probit models

Variables	**Before 1986**				
	Case A	**Case B**		**Case C**	
		O-Pro	**M-Pro**	**O-Pro**	**M-Pro**
Capital amount (log-transformed)	+		+		+
Year of business foundation	+	+	+	+	+
Year of business foundation (squared)	–	+		+	+
R&D	+		+		+
Production	+	+	+		+
Favorable government policies regarding taxes and investment	+		+		+
Favorable government policies regarding trade	–		–	–	–
Transportation infrastructure (roads, ports, etc.)	+	+	+	+	+
Other physical infrastructure (electricity, offices, etc)					
IT infrastructure (broadband, etc.)					
Market located near Bangkok	–		–		–
Market located near Bangkok					–
Suppliers/subcontractors located near Bangkok			+		
Request by large company/related company or govt.					–
Unskilled labor market	+	+	+	+	+
Skilled labor market	–	–	–	–	–
Other companies from the same country are located here	+		+	+	+
Access to cutting-edge information/ technology (spillovers)	–	–	–	–	–
Business field (manufacturing)					
Business field (wholesale)	+	+	+	+	+

Source: author.

After 1987								
Case A	**Case B**				**Case C**			
	1987–'97		**After 1998**		**1987–'94**		**After 1995**	
	O-Pro	**M-Pro**	**O-Pro**	**M-Pro**	**O-Pro**	**M-Pro**	**O-Pro**	**M-Pro**
–		–		–		–		+
–	–	+	–	–	+	+	–	–
+	+		+		+	–	+	+
–		–		+		+		–
–	–	–	–	–		–		–
–	–	–				–		–
+	–	+		–	–	–	–	+
–	–	–	–	–	+	–	–	–
				+				–
+		+		–		+		–
				+				–
		+		–				
						+		–
–	–	–	–	–	+	–	–	+
+	+	+	+	+	–	–	+	+
–		–		–	–	–	–	+
+	+	+	+	+	–	–	+	+
–	–	–	–	–	–	–	–	–

5.7. Estimation results and the flowchart approach

Observing the trends of the Japanese as well as the Thai economies, we find the division of time periods in Model C to be reasonable. Therefore, let us discuss the relationships between the results of Model C and the flowchart approach, We can summarize the similarities as follows:

Anchor firms definitely came first, followed by smaller ones entering the market. After 1995, larger firms began to open offices again.[10] of the first-movers thought of Bangkok as a production base.

Concerning elements of capacity building, "transportation infrastructure such as roads and ports" and "government policy regarding tax and investment" were responsible for attracting first-movers. Human resources are highly evaluated both by first-movers and firms that came after 1995. On the other hand, living conditions are found to be insufficient in Greater Bangkok.

Domestic markets had positive effects on agglomeration during 1987–1994. Export did not have any significance.

Related or affiliated firms mainly came to Bangkok during 1987–1994.

After 1995, "other companies from the same country located in the region" and "access to cutting-edge information/technology" have positive signs. This indicates that after 1995 the density of the agglomeration in Greater Bangkok reached a level sufficient to increase the prevalence of inter-firm activities, as predicted by the flowchart approach.

On the other hand, some results of the estimation do not support the flowchart approach. R&D activity in industrial clusters is considered to be a late-occurring phenomenon, but our estimation shows that R&D has positive signs during the two periods before 1994, and negative signs after 1995. This does not seem to be consistent with the flowchart approach.

According to the discussions above, we can conclude that the process of agglomeration in Greater Bangkok supports most of the core structures of the flowchart approach.

6. Conclusion

We made an attempt to examine the nature of agglomeration in Greater Bangkok by heuristic methodologies, including quantitative analysis based on mail surveys and in-depth interviews. The purpose of the analysis is to examine the verifiability of the flowchart approach with regards to agglomeration in Bangkok. Let us summarize the analysis in two parts.

6.1. Summary of mail survey

Prior to the quantitative analysis, we conducted mail surveys, sending questionnaires to 1,600 firms (including 1,000 Japanese, 300 other foreign, and 300 Thai companies). We received 143 valid responses (117 Japanese, 9 other foreign, and 17 Thai companies). The average response rate was 8.9%

(11.7% for Japanese, 3.0% for other foreign, and 5.7% for Thai companies). By analyzing the mail surveys, we obtained characteristics of agglomeration in the Bangkok area, and by focusing on the "important factors at the time of business establishment" as initial conditions for formation of the industrial cluster in Bangkok, and on "current important factors" as the conditions for future cluster development, we obtained the following results.

Important factors at the time of business establishment:
- Market located near Bangkok
- Transportation infrastructure
- Other physical infrastructure (electricity, offices, etc.)
- Favorable government policies regarding taxes and investment
- Availability of professionals
- Social environment quality (culture, crime rates, etc.)
- Better financial and legal systems

Current important factors
- Living conditions and living costs
- Request by large company/related company or government
- Unskilled labor market

The factors and degree of satisfaction revealed by the companies with Bangkok operations can be listed as follows:

Factors with which respondents are satisfied:
- Living environment
- Markets located near Bangkok
- Social environment quality
- Advantage for export
- Suppliers/subcontractors located near Bangkok

Factors with which respondents are dissatisfied:
- Congestion
- Physical environment quality
- Protection of intellectual property rights
- Legal system

Thus, respondents expressed satisfaction with access to market and suppliers, but dissatisfaction with infrastructure and institutions. Most of the surveyed companies have strong intentions to continue operations in Bangkok; more than half of those intend to increase their investments in Bangkok. In addition to the markets located near Bangkok, these firms are very satisfied with living environment, living costs, and social environment quality. Together with these positive factors, strengthening of human resource cultivation, development of legal systems, and deregulation will help the existing industrial agglomeration in Bangkok to evolve into an industrial cluster that encourages endogenous innovation processes.

6.2. Summary of empirical analysis and flowchart approach

Based on the data obtained from our mail survey, we conducted three types of estimations: Models A, B, and C. These models differ from each other in the division of time period and in estimation methodology. The time periods are determined according to the year business operations were established in Bangkok.

- Model A: two periods: through 1986 and after 1987; binomial probit estimation.
- Model B: three sub-periods: through 1986 (Group 1B); between 1987 and 1997 (Group 2B); and (iii) after 1998 (Group 3B); ordinal multinomial probit estimation.
- Model C: three sub-periods: through 1986 (Group 1C); from 1987 to 1994 (Group 2C); and after 1995 (Group 3C); ordinal multinomial probit estimation.

In Model A, the year 1986 is determined endogenously as one that shows a structural change in agglomeration. Model B chooses year 1997 as its break point, taking the East Asian Economic Crisis as a reference point. In Model C, the year 1995 is selected because it seems to represent a structural shift.

In these three models, we take the years of Bangkok offices establishment as the dependent variables, while attributes of firms and factors leading to agglomeration constitute the explanatory variables. The flowchart approach predicts certain factors as the determinants of locational decision-making; these are domestic demand, export, capacity building such as infrastructure, institutions, human resources, and living conditions. We also take the following factors as explanatory variables: inter-firm relationships such as those between anchor firms and related or affiliated firms, and the degree of agglomeration (such as level of information flow related to technology and other business activities). The most important relationship is the relationship between the year of Bangkok office establishment and firm size. The flowchart approach postulates the hypothesis that anchor firms come first, followed by smaller and related firms. Thus, we make an attempt to prove a negative correlation between these two variables.

The relationships between the estimation results obtained by Model A and the flowchart approach are summarized as follows:

1. Anchor firms are first-movers and smaller ones follow.
2. First-movers thought of the region mainly as a production base.

3. In terms of capacity building, "transportation infrastructure such as roads and ports" and "government policy regarding tax and investment" attracted first-movers. Unskilled labor such as human resources are highly evaluated by first-movers. On the other hand, living conditions were not sufficient.
4. Domestic demand attracted firms during the second period. Export did not have any significance.
5. Related and affiliated firms came to Bangkok during the first period.
6. R&D was a factor driving agglomeration in the first period.

From these results it follows that (1), (2), and (3) are consistent with the flowchart approach, while (4), (5) and (6) are not.

The results of Model B are almost the same as Model A, and this implies that the East Asian Economic Crisis did not significantly affect the patterns of agglomeration in Bangkok.

The relationships between the estimation results obtained by Model C and the flowchart approach are summarized as follows:

1. Anchor firms came first, followed by smaller ones. After 1995, larger firms began opening offices in Bangkok again.
2. In terms of capacity building, "transportation infrastructure such as roads and ports" and "government policy regarding tax and investment" attracted first-movers. Human resources are highly evaluated by first-movers and firms that came after 1995.
3. The domestic market had a positive effect on agglomeration during 1987–1994. Export did not have any significance.
4. Related or affiliated firms mainly came to Bangkok during 1987–1994.
5. After 1995, the density of agglomeration in Greater Bangkok reached a level sufficient for inter-firm activities to increase.
6. R&D has a positive sign during the two periods before 1994, and a negative sign after 1995.

Except for (6), the results of the Model C estimation support the flowchart approach.

6.3. Directions for further research

As discussed previously, the flowchart approach does not explain all agglomeration or clustering phenomena. In Greater Bangkok, the approach is fairly consistent with reality, but not perfect. We analyzed the general situation of agglomeration in Bangkok, taking only the attribute of business fields such as manufacturing and wholesale as dummy variables. Since different industries have different characteristics, we must consider which industrial sector can be better explained by the flowchart approach. Similarly, each country has its own characteristics such as historical background, natural resources, and economic system; thus, agglomeration is not uniform over countries. So far, the flowchart approach seems to be suitable to explain agglomeration in developing counties, especially East Asian economies in which agglomeration is guided and planned by governments. For example, these countries construct special economic zones to attract FDI by offering generous economic benefits. These agglomerations are surely different from, for example, most IT clusters in developed economies. Thus, more international comparison is required in order to expand the verifiability of the flowchart approach.

Regarding international comparative studies, we desire the formalization of the concept of the degree of agglomeration. Each economy is at a different stage of agglomeration; in some counties, foreign anchor firms consider clusters merely as production bases for exploiting abundant local human as well as natural resources. On the other hand, some counties are at the highest stage of clustering, such that R&D activities and innovation processes are formed endogenously. If we succeed in incorporating these R&D and innovation factors into the flowchart approach, it will become a more general theory capable of analyzing most or all phenomena related to industrial clustering.

Appendix

October 2005

JETRO/IDE Research Industrial Clustering: A Case Study of Bangkok

JETRO (Japan External Trade Organization)/IDE (Institute of Developing Economies) has been conducting an international research on industrial clustering in Asia. The research project on Bangkok is conducted with the Sirindhorn International Institute of Technology (SIIT), Thammasat University. We would like to ask managers of businesses operating in Bangkok some questions about choosing business location in the Bangkok region.

Directions

- There are two types of questions in this questionnaire; multiple choice and free response. For multiple choice, please circle the correct answer(s).
- Circling answer(s): (circle one) or (circle all that apply) is noted after some questions. If not specified, please circle one.
- You may be asked to skip some questions. Please follow the instructions.
- **There are some questions that ask about your company's details. The questionnaire's results will be re-arranged statistically and utilized only for academic objects. All information recorded on this questionnaire is strictly confidential.**
- After you finish the questionnaire, please make sure there are no accidental omissions. Please post the questionnaire in the enclosed self-addressed stamped envelop by Monday, **November 7.**
- If you have any questions, please contact:

[Project Director]

JETRO/IDE, Chief of Industrial Cluster Research Group

Graduate School of Applied Informatics, University of Hyogo,

Dr. Masatsugu, TSUJI, Professor

Tel/Fax : +81(0)78–367–8625 Email: tsuji@ai.u-hyogo.ac.jp

Dr. Somrote Komolavanij, Assistant professor
Tel 02–501-3505~20 Email somrote@siit.tu.ac.th

JETRO/IDE Research

Industrial Clustering: A Case Study of Bangkok

- Research Organizations: IDE-JETRO,
- Graduate School of Applied informatics, University of Hyogo
- SIIT, Thammasat University.

- Please write your contact information.

Company name			
Address			
Telephone number		**FAX number**	
Name of responder		**Title/position**	
E-mail			

Section A: Corporate Profile

Q1) Please answer the following questions.

Established Year: Month:	Present CEO is: 1. Founder of the company 2. 2nd CEO 3. 3rd CEO 4. 4th or later	Capital (please specify currency	**Sales of last year (2004)** (please specify currency)

Q2) Your business field(s) (circle one or more)

1. Manufacturing	5. Construction	9. Individual services
2. Wholesale	6. Finance, Insurance	10. Information Technology
3. Retail	7. Real Estate	11. Telecommunications
4. Transportation	8. Corporate services	12. Others ( )

Q2-1) If you checked "1" (manufacturing), what are the manufactured product(s)?

1. Food	9. Metals	14. Other transportation-related machinery
2. Textiles	10. Machinery and tools	15. Chemicals, plastics
3. Woods	11. Computers, computer parts	16. Other ( )
4. Paper	12. Other electronics	
5. Synthetic resin, rubber	13. Cars, auto parts	
6. Ceramic, rock, and sand		
7. Steel		
8. Non-ferrous metals		

Q3) How many employees does your company have, as of October 1st, 2005?

	Total	Regular/full-time	Temporary
Total number			
Number in Bangkok			

Q4) Please tell us which 3 countries (including Thailand) are your company's most important markets and suppliers. For each country, please estimate the percent of sales or purchases that go to/come from that country.

Market Countries and Percent of Sales

1: %	2: %	3: %

Supplier Countries and Percent of Purchases

1: %	2: %	3: %

B Company's history and background in Greater Bangkok

(note: throughout this questionnaire, "Bangkok" refers to Greater Bangkok.)

Q5) When did your company establish its first Bangkok office?

Year:

Q5-1) When did your company first decide to conduct business in Bangkok?

Year:

Q6) Is your company headquartered in Bangkok, or are the Bangkok operations a branch or subsidiary? (circle one)

1 Headquarters
2 Branch
3 Subsidiary

Q6-1) <For Branch or Corporation>Where is/are your headquarter(s) located?

Country	**City**

Q7) What function(s) does your company carry out in Bangkok? (circle one or more)

1	Sales
2	Marketing
3	R&D
4	Production
5	Purchasing
6	Logistics
7	Accounting
8	HR
9	General affairs
10	Other ( )

Q8) How important were each of the following factors in your company's decision to locate its operations in Bangkok, <u>at the time Bangkok operations were begun</u>?

(note: This is the most important question on the questionnaire, so please answer carefully.)

(Please circle one number on each line)

	Very important	Somewhat important	Not sure	Not very important	Not important at all
(1) Favorable government policies regarding taxes and investment	5	4	3	2	1
(2) Favorable government policies regarding trade	5	4	3	2	1
(3) Transportation infrastructure (roads, ports, etc.)	5	4	3	2	1
(4) Other physical infrastructure (electricity, offices, etc.)	5	4	3	2	1
(5) IT infrastructure (broadband, etc.)	5	4	3	2	1
(6) Better financial system	5	4	3	2	1
(7) Better legal system	5	4	3	2	1
(8) Better protection of intellectual property rights	5	4	3	2	1
(9) Market located near Bangkok	5	4	3	2	1
(10) Advantage for export	5	4	3	2	1
(11) Suppliers/subcontractors located near Bangkok	5	4	3	2	1
(12) Request by large company/ related company or govt.	5	4	3	2	1
(13) Unskilled labor market	5	4	3	2	1

Continued

	Very important	Somewhat important	Not sure	Not very important	Not important at all
(14) Skilled labor market	5	4	3	2	1
(15) Availability of professionals	5	4	3	2	1
(16) Other companies from the same country are located here	5	4	3	2	1
(17) Access to cutting-edge information / technology (spillovers)	5	4	3	2	1
(18) Environmental quality (climate, pollution, etc.)	5	4	3	2	1
(19) Social quality (culture, crime rates, etc.)	5	4	3	2	1
(20) Living environment (shopping, entertainment, etc.)	5	4	3	2	1
(21) Rents, land prices, cost of living	5	4	3	2	1
(22) Congestion	5	4	3	2	1
(23) Founder's personal choice (born in Bangkok, etc.)	5	4	3	2	1

1 Very important	3 Not sure	4 Not very important
2 Somewhat important		5 Not important at all

D Current importance of and satisfaction with Bangkok conditions

Q9) How important is Bangkok as a business location for your company? (Circle one)

Q10) How satisfied are you with Bangkok as a business location? (Circle one)

1 Very satisfied	3 Not sure	4 Not very satisfied
2 somewhat satisfied		5 Not satisfied at all

Q11) How important are the following factors in your company's decision to continue/expand its Bangkok operations? (Circle one on each line)

And how satisfied are you with the current condition of each of these factors in Bangkok? (Circle one on each line)

	Very important	Somewhat important	Not sure	Not very inportant	Not important at all	Very satisfied	Somewhat satisfied	Not sure	Not very satisfied	Not satisfied at all
(1) Favorable government policies regarding taxes and investment	5	4	3	2	1	5	4	3	2	1
(2) Favorable government policies regarding trade	5	4	3	2	1	5	4	3	2	1
(3) Transportation infrastructure (roads, ports, etc.)	5	4	3	2	1	5	4	3	2	1
(4) Other physical infrastructure(electricity, offices, etc.)	5	4	3	2	1	5	4	3	2	1
(5) IT infrastructure (broadband, etc.)	5	4	3	2	1	5	4	3	2	1
(6) Better financial system	5	4	3	2	1	5	4	3	2	1
(7) Better legal system	5	4	3	2	1	5	4	3	2	1
(8) Better protection of intellectual property rights	5	4	3	2	1	5	4	3	2	1
(9) Market located near Bangkok	5	4	3	2	1	5	4	3	2	1

Continued

	Very important	Somewhat important	Not sure	Not very important	Not important at all	Very satisfied	Somewhat satisfied	Not sure	Not very satisfied	Not satisfied at all
(10) Advantage for export	5	4	3	2	1	5	4	3	2	1
(11) Suppliers/subcontractors located near Bangkok	5	4	3	2	1	5	4	3	2	1
(12) Request by large company/related company or govt.	5	4	3	2	1	5	4	3	2	1
(13) Unskilled labor market	5	4	3	2	1	5	4	3	2	1
(14) Skilled labor market	5	4	3	2	1	5	4	3	2	1
(15) Availability of professionals	5	4	3	2	1	5	4	3	2	1
(16) Other companies from the same country are located here	5	4	3	2	1	5	4	3	2	1
(17) Access to cutting-edge information / technology (spillovers)	5	4	3	2	1	5	4	3	2	1
(18) Environmental quality (climate, pollution, etc.)	5	4	3	2	1	5	4	3	2	1
(19) Social quality (culture, crime rates, etc.)	5	4	3	2	1	5	4	3	2	1
(20) Living environment (shopping, entertainment, etc)	5	4	3	2	1	5	4	3	2	1
(21) Rents, land prices, cost of living	5	4	3	2	1	5	4	3	2	1
(22) Congestion	5	4	3	2	1	5	4	3	2	1
(23) Founder's personal choice (born in Bangkok, etc.)	5	4	3	2	1	5	4	3	2	1

Q12) Do you have any other comments about points of importance or satisfaction regarding Bangkok's business conditions?

Points of Importance (for continuing business in Bangkok)
Points of satisfaction/dissatisfaction

Q13) Does doing business in Bangkok present problems to your company? (Circle one)

1 Yes 2 A little	3 Not sure	4 Not much 5 Not at all

Q13-1) Please tell us briefly the most important problem(s) your company has faced when doing business in Bangkok.

E Future plans

Q14) Will your company continue its business in Bangkok?

1. Yes	2. Probably	3. Not sure	4. Probably not	5. Not at all

Q14-1)<If you answered 1 or 2 on Q14> Will your company increase its investment in Bangkok?

1. Very much	2. Somewhat	3. Not sure	4. Not much	5. Not at all

Q14-2)<If you answered 4 or 5 on Q14>Please tell us the reason(s). (Circle all that apply)

1. Head office's order
2. Change in government regulation
3. Market change
4. Supplier change
5. Major customer change
6. Language problems
7. Environment (including social environment)
8. Rent and land
9. Living costs
10. Better location has been found
11. Other ( )

Q15) Do you have plans to start new operations somewhere else in Thailand?
If Yes, please write the name of the city.

1. Yes	2.	Not sure	3. No

City's name

Q16) Do you have plans to start new operations in counties other than Thailand?
If Yes, please write the name of the country and city.

1. Yes	2. Not sure	3. No

Country's name	City's name

F Government policies

Q17) What Thai government policies would help your company do business in Bangkok?

Tax cuts / tax incentives
Deregulation
Lower interest rates
Lower lease prices
Subsidies
Improvements in education
R&D improvement
IT improvement
Office environment improvement
Congestion relief
Other ( )

Q17-1) Please tell us your three most important policy requests (if applicable).

First	Second	Third

This is the end of the questionnaire. We really appreciate your cooperation.
Please make sure you have no accidental omissions.
Then please post this questionnaire in the enclosed self-addressed stamped envelop by November 7, Monday.

Notes

1. Sample firms were selected from the member list of the Chamber of Commerce of Japanese Firms in Bangkok for Japanese firms. Foreign firms were selected from the list provided by the BOI (Board of Investment), the Government of Thailand (CD-ROM).
2. This phenomenon is also found in the Guangzhou Area in China, for instance. Japanese firms especially tend to purchase business support services from other Japanese firms, because of the relationship among headquarters in Japan, quality of services, etc. ; see Tsuji and Wu (2005), Kuchiki (2005).
3. Because of this, Bangkok is referred to as the "Detroit of Asia". All Japanese automobile assemblers have been expanding their production capabilities, and more firms in this field are expected to agglomerate in the Greater Bangkok area.
4. Multinational companies cannot wait for the results of official training programs, and thus they themselves, their consortiums, and Japanese organizations such as JETRO have started their own training programs for nurturing the skill of Thai workers as well as students.
5. Regarding labor markets, we asked the following three questions: (13) Unskilled labor market, (14) Skilled labor market, and (15) Availability of professionals. It is obvious that the latter two have a strong correlation, so (15) was omitted from our estimations.
6. These two are selected because they have large enough numbers of replies.
7. Toyota Motor Corporation established an R&D center in Bangkok, which is the firm's third largest in the world, after the centers in Japan and the U.S. This is a symbol that agglomeration in Greater Bangkok has entered a new, advanced stage.
8. The sum of marginal effects regarding one explanatory variable for each group is zero.
9. In order to complement this, we examined a probit model, adding a dummy variable for the Asian financial crisis and grouping the business foundation years in the same manner as Case A (one group for Bangkok operations begun after 1987). However, we could not obtain convergent values. This is because there might be a variable that would be zero or one for all replies of one group.
10. The behavior of large firms that entered the region can be interpreted as follows: large firms such as petrol-chemical firms that require huge amounts of investment for establishing production bases may wait for investment until domestic demand in a host county becomes large enough.

Bibliography

Fujita, Masahisa and Jacques-François Thisse, *Economics of Agglomeration: Cities, Industrial Location, and Regional Growth*, Cambridge, UK and New York: Cambridge University Press, 2002.

Fujita, Masahisa, Paul Krugman, and Anthony J. Venables, *The Spatial Economy: Cities, Regions, and International Trade*, Cambridge, MA: The MIT Press, 1999.

Greene,W. H., *Econometric Analysis*, 4th edition, Upper Saddle River, N.J.: Prentice-Hall, 2000.

Kuchiki, Akifumi, "A Flowchart Approach" in Kuchiki and Tsuji (2005), pp. 169–99, 2005.

Kuchiki, Akifumi, "Agglomeration of Exporting firms in Industrial Zones in Northern Vietnam: Players and Institutions," in Tsuji, Giovannetti and Kagami (2007), pp. 97–138, 2007.

Kuchiki, Akifumi and Masatsugu Tsuji, *Industrial Clusters in Asia: Analyses of their Competition and Cooperation*, Basingstoke, UK: Macmillan, 2005.

Krugman, Paul, *Development, Geography, and Economic Theory*, Cambridge, MA: The MIT Press, 1995.

Tsuji, Masatsugu, Shoichi Miyahara and Makoto Ishikawa, "An Empirical Analysis of Industrial Transformation in the Japanese Machine Tool Industry," *Mathematics and Computers in Simulations*, Vol. 48, No. 4–6, pp. 561–72, 1999.

Tsuji, Masatsugu, Emauele Giovannetti and Mitsuhiro Kagami, *Industrial Agglomeration and New Technologies*, Cheltenham, UK: Edward Elgar, 2007.

Tsuji, Masatsugu and Wu Quan, "Chinese Automotive and Parts Industries," in Kuchiki, A. and M. Tsuji (eds.), *Industrial Clusters in Asia: Analyses of Thier Competition and Cooperation*, London: Macmillan, 2005, pp. 220–21.

8

The Evolution of the High-Tech Electronics Cluster in Guadalajara, Mexico

Yoshiaki Hisamatsu

1. Introduction

This chapter examines how and why Guadalajara, the second biggest city in Mexico, has risen to prominence as "the Mexican Silicon Valley." In so doing, it also scrutinizes various key factors in cluster development, applying the framework of the "flowchart approach to industrial cluster policy." The rest of the chapter is structured as follows. The second section shows the export and investment performance of the cluster, and explains the basic structure and principal enterprises present in the cluster. The third section examines why and how major companies chose the Guadalajara area as a production site. The fourth section explores the challenges faced by the cluster from the late 1990s onward. With the end of the information technology bubble in the U.S. and the rise of China, the cluster is faced with formidable challenges. The fifth section examines local linkages within the cluster. The sixth section examines the several elements of the "flowchart approach" for case of the Mexican Silicon Valley. The seventh section concludes the chapter.

2. Performance, actors, and structure of the Mexican silicon valley

The most outstanding feature of the electronics industry in Guadalajara and its surrounding area in the state of Jalisco is its vigorous export performance (Figure 8.1). In 2005, the value of exports of the Jalisco electronics industry was US $11.3 billion, representing more than 70 percent of the state's total exports. This figure indicates the importance of the sector in the local economy. A more notable fact is that exports increased 6-fold in the 6 years between 1994 and 2000.

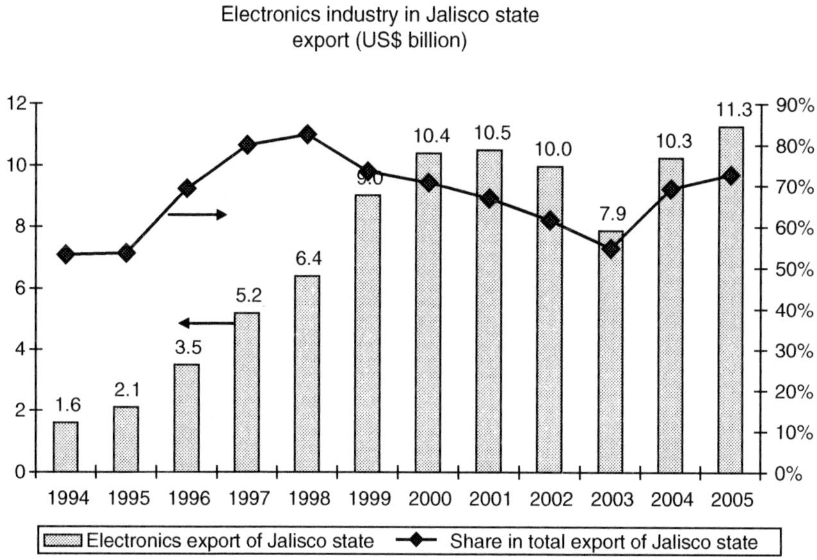

Figure 8.1 Electronics exports from Jalisco state

Note: The 2005 number is an estimate.

Source: CADENA PRODUCTIVA DE LA ELECTRÓNICA, A.C.

This impressive export record was achieved through massive investment made by foreign (mostly American) electronics companies (Figure 8.2).[1] This massive investment continued through 1999, but since 2000 investment has moderated considerably.

This electronics cluster consists mainly of multinational companies. The major company list includes household-name companies like IBM, Hewlett Packard (HP) and Intel (Table 8.1). This list has four rows. The first two rows are occupied by Brand Companies (BCs), which have their own brands and marketing functions. The last two rows are occupied by Contract Equipment Manufacturers (CEMs) and/or Electronic Manufacturing Services (EMSs), which carry out the role of production in the supply chain. In addition to BCs and CEMs, the cluster includes a number of design centers and numerous specialized suppliers.

The various kinds of companies described above form an industrial cluster (Figure 8.3). Both specialized suppliers and design centers provide BCs and CEMs with their products. BCs and CEMs assemble parts into products and export most of those products to the U.S. Therefore, with regard to the supply chain, the Mexican Silicon Valley functions largely as manufacturing base. This is the major difference between Guadalajara and the original Silicon Valley, which functions principally as a research and development (R&D) center.[2]

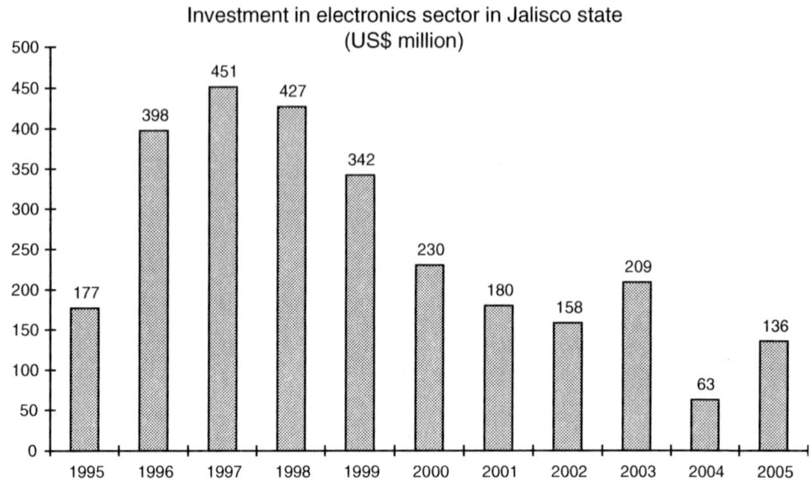

Figure 8.2 Investment in electronics in Jalisco state

Note: The 2005 number is investment made up to August.

Source: CADENA PRODUCTIVA DE LA ELECTRÓNICA, A.C.

Table 8.1 Major companies in the Guadalajara electronics cluster

Function	country	Company name
BCs	International	IBM, HP, Intel, Siemens VDO, Hitachi Global Storage Technologies, Kodak, Technicolor, BDT
	National	Resser, ATR, Industrias IDEAR, Pegasus Control
CEMs / EMSs	International	Solectron, Sanmina-SCI, Flextronics, Jabil Circuit, USI, Benchmark, Mexico Scientific, Telect, Mexikor, Vogt Electronics, Foxconn, Yamaver-EPIQ
	National	Logan MEX, SERIIE, EEJ

Source: CADENA PRODUCTIVA DE LA ELECTRÓNICA, A.C.

This cluster manufactures various kinds of high-technology electronic products and components. Numerous examples include personal computers, servers, robotic tape libraries, storage systems, set top boxes, photo compact discs and digital video discs, jukeboxes, relays, vending machines, hubs, internet protocol phones, printers, telecom servers, routers, firewalls, docking stations, handhelds, medical equipment, mobile phones, cameras, car immobilizers, car security systems, air bag systems, anti-lock braking systems, paper handling solutions, internet access boxes for television etc.

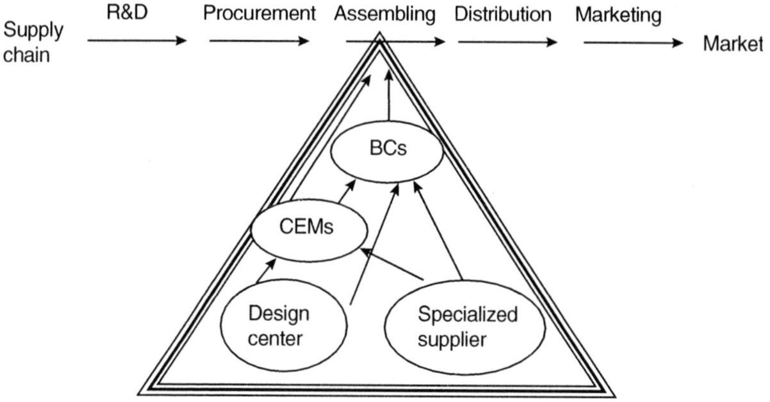

Figure 8.3 Structure of the Guadalajara electronics cluster
Source: author.

3. The locational advantages of the Mexican silicon valley

As the previous section clearly shows, this cluster is a high-tech production center for multinational corporations. This section explains why and how Guadalajara and its surrounding area evolved into Mexico's high-tech electronics industry center.

There are various factors that help explain why foreign electronics companies chose Guadalajara. It is apparent that the location was selected based on a mix of a number of external and internal factors. As an industry, electronics demands a high degree of flexibility, which is suggested by the emergence of several contract electronics manufacturers (CEMs). This flexibility requirement favors firms with location near the market, and Guadalajara, accessible with good highways, is close to the U.S. market. The industry also requires abundant human resources. Guadalajara is adequate in this regard, as evidenced by the existence of 10 universities in the city.

3.1. Electronics industry trends

Since the late 1980s, many Brand Companies (BCs) in the electronics industry have been increasingly utilizing Contract Electronics Manufacturers (CEMs) or Electronics Manufacturing Service providers (EMSs) in their business and manufacturing strategies, and are seeking to outsource a broad range of manufacturing and related engineering services. Outsourcing allows BCs to take advantage of the manufacturing expertise and capital investments of CEMs, thereby enabling BCs to concentrate on their core competencies, such as research and development, marketing, and sales. BCs utilize electronics CEMs to enhance their competitive position by:

- reducing production costs;
- accelerating time-to-market and time-to-volume production;
- accessing advanced manufacturing and design capabilities;
- reducing capital investment requirements and fixed overhead costs;
- improving inventory management and purchasing power; and
- accessing worldwide manufacturing capabilities.

As a result, industry sources estimate that the overall market for EMS grew at an average annual rate of about 25 percent from 1988 to 1998, reaching an estimated USD $90 billion in 1998. The top 10 CEMs have established an extensive network of manufacturing facilities in the global major electronics markets - Asia, the Americas, and Europe - to serve the increased outsourcing needs of both multinational and regional BCs. The reason why the Guadalajara area is exaggeratedly called "the Mexican Silicon Valley" is that four of the top five BCs have established their factories in the area.[3] This fact means that Guadalajara area has become a high-tech cluster comparable to Penang, Malaysia; Guangdong, China; and Dublin, Ireland. All the BCs and CEMs locate their R&D facilities in the U.S., mostly in the real Silicon Valley.

The outstanding expansion of CEMs was achieved not only by "greenfield investments," but also by acquisitions of BCs' factories. In another words, major BC companies like IBM, Ericsson, Hewlett Packard, Cisco etc. have sold their own manufacturing facilities to CEMs to focus on research and development, quality control, and marketing and distribution. Computer makers like Dell and Gateway do not have to own so many manufacturing facilities, and obtain their products through contract with CEMs. On the other hand, CEMs have expanded their operating scale to reduce the cost of production. The fact that many BCs still have some manufacturing facilities, however, shows that CEMs compete not only among themselves but also against BCs' internal production.

CEMs basically produce two kinds of products: key parts such as Printed Circuit Boards (PCB), and final products. Supply chains are different, depending on whether CEMs produce key parts or final products. In the former case, CEMs only deliver key parts to BCs, which manufacture the final products. In the latter case, CEMs produce final products and deliver them to BCs' distribution channels. In this second case, one might see, in a CEM's factory, one production line for Philips Web TVs and another production line for Apple modems. It is said that doing final product business is more profitable than making only key parts. Key parts are outsourced because they are more standardized products with lower margins than final products.

The reason BCs are relying more and more on CEMs for manufacturing is the rapidly changing environment in the electronics industry, which calls for ever greater specialization.[4] In competitive segments of the industry, it is not unusual for new products to come onto the market every 3 months with lower prices and more features. In this kind of competitive environment, BCs want to focus on their core businesses. Therefore, BCs concentrate their attentions on R&D, quality control, marketing, and distribution, while CEMs specialize in manufacturing. Another concern in a rapidly changing environment is inventory management.

Inventory holding is very costly when new products emerge every 3 months. Therefore, BCs naturally avoid holding inventory. This is why supply chain management (SCM) has become common throughout the industry. In the electronics industry, BCs usually rely on the "just-in-time" production system. In addition, a BC (for example, IBM) may provide suppliers with factory space for manufacturing, only recognizing final products as its own after they are thoroughly tested. This means that products in process and/or final products in testing count as suppliers' inventories. If some parts prove defective in testing, the failure belongs to the supplier, not to the BCs. This arrangement requires suppliers (CEMs and specialized suppliers) to maintain high-quality management and just-in-time delivery. When CEMs manufacture goods, BCs do not have to worry about inventories. BCs need be concerned only with their own inventories at the distribution center. When they do not want to accumulate additional inventories, they merely inform CEMs of this fact; CEMs must then worry about their own inventories. CEMs also set up their own inventory management system, in which their own suppliers have to hold inventories in the "just-in-time" production system. CEMs claim that their specialized skill is manufacturing, not holding inventories.[5] As a result, all actors including specialized suppliers involved would manage inventories efficiently. Because of the above-mentioned factors, CEMs strategically locate their facilities near their customers' end markets and have located fully integrated, high-volume manufacturing facilities in low-cost regions worldwide. By operating in low-cost areas, they are able to realize savings through lower labor, overhead, tax and transportation costs, which they can pass on to their customers.

In summary, the electronics industry chooses manufacturing locations with the following characteristics:

- Low-cost structure
- Market proximity
- Relatively abundant high-quality human resources in engineering for quality management
- Any other factors that promote flexibility

3.2. Guadalajara's locational advantages

There are six factors that may explain the locational advantages of Guadalajara area. First, note once more that from 1994 onward electronics exports from the Mexican Silicon Valley increased dramatically (Figure 8.1). It is certain that export growth was closely related to Mexico's currency devaluation in late 1994. This devaluation drove down labor costs in all Mexican cities. All of a sudden, Mexico had low-cost labor; as a result, CEMs started investing in Mexico. Another low-cost factor is the North American Free Trade Agreement (NAFTA), combined with Maquiladora and other export promotion arrangements of the Mexican government.[6] In the Maquiladora setup, companies could import inputs for export without tariffs. The United States has another arrangement that only imposes tariffs on value-added components from Mexico. This institutional combination strongly favored Mexico as a low-cost manufacturing site for the U.S. market.[7]

The second factor is market proximity. Mexico is one of only two countries that have a long border with the United States.[8] This market proximity favors not only one but almost all border cities in Mexico, such as Tijuana, Mexicali, Cd. Juarez, etc. Guadalajara is not geographically a border town, but it is important to note that in the industrial world, the quality of transportation facilities also defines market proximity. Currently, the most convenient gate between the U.S. and Mexico is the Nuevo Laredo (Tamaulipas state, Mexico) – Laredo (Texas state, U.S.) border. The fact that there is a good highway connecting Laredo, Monterrey, and Guadalajara favors both Monterrey and Guadalajara, along with the border cities, from a market proximity point of view.

The third factor is also specific to high-technology production. High-technology production involves some learning processes that require stable labor market environments. Given the high labor turn-over in border towns, companies are inclined to establish factories in areas where rotation is more stable.[9] In this respect, border towns are unfavorable because workers might cross the border anytime.

Fourth, high-technology electronics industries need cities with large quantities of high-quality human resources in engineering.[10] Since quality management is vital in high technology, these industries hire more engineers than traditional electronics industries (audio and television).[11] Both Monterrey and Guadalajara have quite a few technology-oriented universities.[12] In this respect, border cities are not as competitive as big cities like Monterrey and Guadalajara.

The fifth and sixth factors favor Guadalajara against Monterrey. On the one hand, a few BCs such as IBM, Hewlett Packard and Motorola have long operated in Guadalajara, beginning their operational restructuring efforts in the late 1980s.[13] This prior investment encouraged many CEMs and parts suppliers to establish their operations near large BCs already located in

Guadalajara. In addition, the Jalisco state government promotes the establishment of the electronics industry. In 1995, the business-minded opposition party (PAN) won the state governorship, and former businessmen took state cabinet posts. The state government enhanced favorable tax treatment, established a single-window processing system (meaning companies can deal with all bureaucratic procedures at a single window), and built infrastructure support.

In sum, the emergence of the Mexican Silicon Valley is explained by various factors such a s macro-economic contingency (devaluation), market proximity, labor market environment, human resources, history (prior investment), and institutional support.

3.3. Navigators in cluster development: the role of local professional managers

There is a need, however, to transform Guadalajara's advantages into a well-articulated and coherent locational competitive advantage. Deliberate and coordinated actions support specific objectives, such as forming a layer of local suppliers and service providers, and promoting institutions in fields like technology information and extension or training. These efforts tend to strengthen local specialization profiles and create locational competitive advantages. Coordination requires the cooperation of all these actors. In the Guadalajara electronics sector, local professional managers do take collective initiatives through their business associations.

3.3.1. The rise of local professional managers

Many companies in the Guadalajara electronics cluster are affiliates of multinational companies, and in many cases local professionals fill the top positions in these companies. This means that the top local manager is not the owner of the company; a stark contrast to many big companies in other developing countries, where the owners are frequently top managers. Family businesses, which may afford management stability in legally fragile economies, are very typical in big companies in most developing countries.

Like other emerging economies, Mexico used to have three types of enterprises: public enterprises, multinational enterprises, and domestic private enterprises. The structural adjustment in 1980–90s decreased the number of public enterprises substantially through privatization, even though public enterprises remain dominant in the electricity and petroleum sectors. Many multinational companies are listed in foreign stock markets, and many domestic enterprises are still managed by their owners in Mexico. In sum, the separation of management and ownership in domestic private enterprises in emerging markets is not so frequent as in developed markets.[14]

A big defect of owner management lies in the lack of sector-specific skills. When an industry evolves steadily and slowly, an owner can eventually learn to become a talented manager. In the rapidly evolving and changing high-tech electronics sector, however, there is a need for professional managers. This is a big obstacle for domestic family businesses to enter the electronics sector. Partly as a result, Guadalajara's electronics cluster started with almost entirely multinational companies. These companies hire professional managers locally. The most important specific skill of the local professional managers is to adapt to the rapidly changing environment of the electronics sector. Since Guadalajara is largely a manufacturing site, the job requires certain engineering knowledge, but the requirement in this highly evolving field includes more than just that knowledge. Whenever a new product that demands a new form of production comes out, professional managers need to learn quickly and adjust themselves very rapidly, not only in engineering knowledge but also in management skills. Managers are under extreme pressure by U.S. corporate headquarters to attain certain performance standards; this is especially the case when the products are headed for the U.S. market.

Given the initial conditions from Mexico's import-substitution era, the Guadalajara electronics cluster has nurtured and retained a pool of local professional managers, who have become knowledgeable in engineering, flexible to production changes, and well-trained to deal directly with local workers. Since these managers have developed their specific skills locally, they naturally have a stake in growing the cluster.[15]

3.3.2. Business association composed of local professional managers

Local professional managers in the Guadalajara cluster use their business association to coordinate industry development promotion.

In general, business associations in developing countries function as a kind of social club as well as a mouthpiece for communication with the political hierarchy – in other words, a coordination device for demanding something from the outside.[16] The general Mexican case is not an exception. Companies in Mexico are obliged to affiliate with a business association. This has much to do with the corporatist structure of the Mexican political economy. In 1993, the reform of the Business Association Law abolished the obligatory affiliation of business association; however, the fact that companies must register in the Mexican Enterprise Information System (Sistema de Información Empresarial Mexicano: SIEM), and that a business association manages SIEM, makes company affiliation with a business association de-facto necessary. Thus, business associations typically function as interest groups.

The Delegation of the Occidental Region of the National Chamber of the Electronics, Telecommunications and Information Industry (Cámara Nacional de la Industria Electrónica, Telecomunicaciones e Informática, Delegación Regional Occidente: CANIETI-Occidente) also functions as an interest group. However, it has two unique aspects that foster an additional function: horizontal coordination efforts within the sector.

The first of these aspects is that the managers, not the owners, of companies are the main members of this business association. As described above, the electronics industry in Mexico is largely occupied by multinational enterprises, which means that local professional managers, not owners, are the main members of the business association. This membership profile influences incentive structures within the business association. Professional managers have specific skills of business management largely particular to one industry. In addition, local professional managers have personally invested their skill in a specific industry, compared to local owners who can invest their assets into different industries. In other words, professional managers have a strong stake in the development not only of their companies but also of their industry.[17] Consequently, they are more interested in receiving benefits from coordinating actions to promote their cluster. The headquarters of multinational companies constantly monitor their managers' activities. If the managers' efforts are regarded as collusive, those managers might lose their jobs. But, the headquarters themselves might favor such cooperative efforts of local professional managers, as long as the efficiency of subsidiaries is enhanced. When a cluster becomes multi-layered, the efficiency of inventory management increases due to external economies; that is why the global trends of the electronics sector favor creating multi-layered clusters. As a result, in the case of the Guadalajara electronics sector, local professional managers are able to cooperate with each other to promote such a cluster. For example, they have used their business association to launch jointly an initiative to invite suppliers to Guadalajara from abroad and promote related sub-industries such as software or design centers.

The second feature of CANIETI-Occidente that favors horizontal integration has to do with government institutions. The National Chamber of the Electronics, Telecommunications, and Information Industry allows each regional delegation to conduct its own independent function and mission. Currently, there are 6 regional delegations in the country (Tijuana, Ciudad Juarez, Aguascalientes, Mexico City, Monterey, and Guadalajara). Each delegation serves its members' interests. This decentralized structure allows CANIETI-Occidente to conduct its own locational coordination efforts.

3.3.3. Spin-offs of specialized organizations

A distinguishing feature of CANIETI-Occidente is its spin-off movement (Figure 8.4). During the rapid expansion of the electronics cluster in Guadalajara, local professional managers had recognized that local suppliers were necessary to expand their business and reduce production costs. The important fact is that this objective is consistent with both these individuals' incentives as professional managers and their incentives as business association members. Since local suppliers are employed on a contract basis, a local supplier does not entirely depend on one company.[18] Given this, all

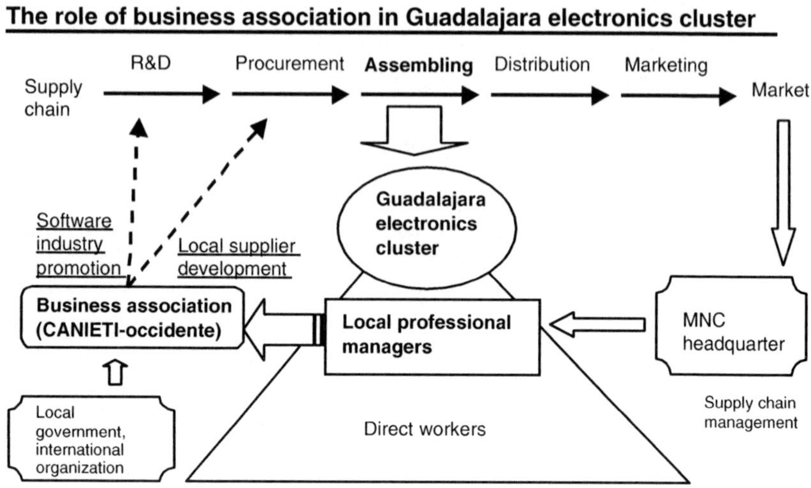

Figure 8.4 The role of the business association in Guadalajara
Source: author.

the local professional managers have a collective interest in having a local supplier market. Therefore, in the mid-1990s, CANIETI's vice-presidency for local supplier development required restructuring. Faced with increasing demand for its activities, the organization decided to spin these activities off into a separate entity. That is how the civil association called Electronics Productive Chain (Cadena Productiva de la Electrónica, A.C.: CADELEC) was formed in 1998 with the support of the state government, the United Nations Development Program (UNDP), and FUNTEC, a technology-transfer-oriented foundation. Through close coordination with CADELEC, CANIETI-Occidente can still maintain consistent sector-wide coordination while delegating specialized activities to this special organization.

CADELEC provides specific services to companies interested in attracting local suppliers. It surveys host companies regarding their specific local supply needs, and maintains a database of this information. For a specific niche market of local supplies, CADELEC conducts market research and sells the research data on a cost-incurred basis.

A typical methodology is as follows. For a certain product, CADELEC makes a questionnaire survey and distributes it to companies that use that product. A typical questionnaire asks whether companies use a certain product or not; whether user companies want to use local suppliers or not; how much the total demand might be; which companies use the product more; and the names and contact addresses of potential clients. Using the questionnaire, CADELEC creates a report on a particular product market. The report can identify an overall demand of the Guadalajara cluster for a particular product, rather than only a single company's demand. The impartiality of CADELEC allows it to do credible research efficiently, which is why all Guadalajara companies cooperate in the organization's efforts. In addition, CADELEC is engaged in research on topics of concern to the cluster, such as the labor market environment.[19] In sum, CADELEC is a nonprofit organization that is able improve information asymmetry between the possible demand side (BCs and CEMs) and the supply side (specialized suppliers). Results have so far been positive.

This spin-off attempt of CANIETI-Occidente is not exceptional. Recently, local professional managers have been realizing that manufacturing is not enough to consolidate the electronics business, and that it is time to move on. After they had participated in a best practice tour to Canada, India, and Ireland in 1999, Guadalajara's managers decided to promote local R&D, such as design of hardware and software. That is how the Jalisco Information Technology Institute (Instituto Jalisciense de Tecnologías de la Información: IJALTI) was created in 2002, again as a civil association with the support of the state government. Notably, this is a joint initiative with principal universities in the area. However, the results of JALTI as of 2005 have been less than expected at the time of initiation.

4. Fragile advantages and challenges: cluster development after 2000

4.1. The Mexican silicon valley in crisis

As indicated above, the Guadalajara area had locational advantages, which were the engines that first pushed drastic export growth. But these were and still are rather fragile advantages. Since the year 2000, the cluster has been faced with a number of challenges. These include the termination of the Maquiladora arrangement at the end of 2000, exchange rate appreciation due to the recovery of the Mexican economy, the end of the IT bubble in the U.S., and the emergence of China as competitor.

The termination of the Maquiladora arrangement from the year 2001 imposed on manufacturing companies Mexico's standard tariff system, under which they must pay tariffs on non-NAFTA input imports according to the tariff schedule. In the Maquiladora arrangement, manufacturing companies were only required to pay tariffs on the Mexican value added. This is a very important issue for the electronics industry in the Guadalajara area because of two factors. First, electronics manufacturing there depends heavily on imported inputs – about 90 percent of total parts are imported from non-NAFTA countries. Second, the cost portion of human resources in manufacturing service is quite low – less than 5 percent of total costs. The combination of these two factors places CEMs in a very tariff-sensitive position. The locational advantage of Guadalajara is thus very precarious, depending on the tariff rate that Mexico now imposes on imported inputs.

The Mexican government has introduced an alternative tariff-reducing scheme called PROSEC (Programas de Promoción Sectorial) to alleviate the situation.[20]

In the above discussion, we implicitly assume that CEMs can move their production facilities from Mexico to Asia. In another words, land and building costs are cheap or can be sold at or near the purchase price, and equipment is movable. Investment must be reversible. According to our interviews, we believe that this is probably the case. Land and building costs were cheap at the time of purchase, partly because of preferential treatment. Equipment is very much movable by air and/or maritime transportation.

Actually, after 2000 there was some factory closure and equipment movement.[21] There were four reasons for this. First, Mexico modified its tariff arrangement as discussed above. Second, Mexico's economic recovery resulted in a gradual appreciation of the exchange rate, reducing Mexico's advantages as a production site. Third, the burst of the IT bubble in the U.S. and the subsequent economic downturn reduced demand in the sector. Finally, the rise of China as a competitor changed the situation of global relocation of production.

The end of the IT bubble and the rise of China are related not only in labor cost terms, but also in terms of the importance of market proximity. The end of the IT bubble certainly increased product cycles. Under a longer product cycle, longer product transport time is not as problematic because products don't become obsolete in the meantime. Companies spend more time introducing and delivering new products to the market. Thus, the importance of market proximity decreases. Because of these factors, exports stumbled after 2001 (Figure 8.1).

4.2. Ongoing challenges for the Mexican silicon valley

The value of this cluster's exports set a record in 2005 for the first time in five years (Figure 8.1). The economic recovery and boom of the U.S. clearly has exerted a favorable influence on the cluster. However, the industry is in a state of constant reshuffle. It is expected that scale-economy manufacturing will moves to and stay in China. Many local managers consider this as a matter of fact, and are heading for production of more value-added goods, establishing design centers and hiring software engineers. HP has set up telephone call centers in Guadalajara catering for Hispanics in the U.S. Many BCs are upgrading technology and carrying out manufacturing processes that used to be done in the U.S.[22] New products include auto parts, medical equipment, and aircraft components.

A notable feature of recent trends is the long-anticipated emergence of local-capital companies. These began in the cluster as packaging or other indirect services, and recently have been involved in the sector as specialized suppliers.[23] This is good news for the cluster, as evidence that it is supported not only by its own human resources but also by its local capital.

However, whether this represents a general trend or is simply a few exceptional cases remains to be seen.

5. Local linkages

This section examines local linkages between multinational electronics companies in the Guadalajara area. Public policy makers in all high-tech clusters have been faced with this problem. The section deals with the issues of outsourcing, obtaining credit and financial services locally, and human resource issues.

5.1. Weak but growing local outsourcing

According to our interviews, less than 10 percent of total parts used in this cluster are procured locally. Most local providers are actually of multinational company origin. Some suppliers were invited from abroad by BCs and/or CEMs, while other multinational supplier companies invested locally on their own. Locally owned and operated companies are emerging, but are still a minor part of the cluster (Table 8.1).

Our interviews confirm that this small local presence is due to three factors. First, both quality and production standards are very high, while local technology, though improving, is still not entirely sufficient. Second, obtaining credit is an obstacle. Historically high interest rates in Mexico apparently dissuade local potential suppliers from investing in the necessary equipment and expansion. Another financial difficulty is that both BCs and CEMs try to minimize inventory costs, forcing parts suppliers to absorb these costs. Given the historically high and volatile domestic interest rate, local suppliers are squeezed financially. Third, even though there are potential investors in the local market, they are generally very cautious. From 1976 to 1994, an economic crisis emerged every 6 years, coinciding with presidential elections. In this volatile macro-economic environment, history-conscious local investors have many reasons for hesitating to invest.

5.2. Weak local financial linkages

Our interviews confirm that both international BCs and CEMs, able to tap the U.S. financial market for any funding needs, do not usually go to the local banking sector for loans. This is not to say that these firms do not use local banking facilities. They turn to Mexico-located banks in two ways: to hold US dollar deposits required for making foreign payments, and to pay local employees. Foreign banks with local branches and major domestic banks are used for the first purpose, and Guadalajara-oriented banks with many automatic teller machines and local branches are used for the second. These operations are very limited. The effect of export-driven cluster growth on banking recovery though not zero is very small. Local banks provide local parts suppliers with some financial arrangements, but the size of these is very limited.

5.3. Promising human resource linkages

We have noted that one of the advantages of the Guadalajara area is its endowment of engineers, technicians, and managers from local universities. We stress that this is an important point of contrast between the Guadalajara electronics export industry and border Maquiladora industries. According to our interviews, the high-tech electronics industry seems to require better-trained human resources than does the traditional electronics industry. CADELEC (2005) shows that the shares of direct and indirect workers are 70 percent and 30 percent, respectively.

Maquiladora industry statistics are consistent with this view. In 2004 the ratio of direct to indirect workers in border states (Baja California, Chihuahua, Coahuila, Sonora, and Tamaulipas) was 79 to 21, compared to 69 to 31 in Jalisco (see INEGI (2005)). Jalisco state hires slightly more indirect workers than border states do.

These differences arise because, even in manufacturing, high-tech industries need higher quality management than other industries. For example, many BCs and CEMs in the Guadalajara area have obtained ISO 9000 and 9001 certifications.[24] In order to do so, they needed capable engineers and local administrators. Also, the industry trend is to require higher quality human resources at manufacturing sites; as a result, high-tech manufacturing clusters demand not only low-cost labor but also high-quality engineers. In this regard, CADELEC sets up arrangements to make local technological universities responsive to industry needs. One CEM company itself has established an internship institution with local universities for students to be ready to start working when they graduate from universities. An BC company designs its factory to imitate college, in order to enhance learning activity. Universities have introduced new academic programs and research centers to satisfy the needs of firms in the cluster (see Palacios (2005)).

We have examined three linkages present in the high-tech electronics cluster of the Guadalajara area. In sum: local outsourcing is a time-consuming process; the impact on the financial sector is weak; and there has been a big impact on human resource development. However, these linkages require time to bear fruit. There is no quick fix for Guadalajara's problems.

6. The Mexican silicon valley examined using the flowchart approach

6.1. The flowchart approach

This approach uses a flowchart to record the evolution of regional industrial development and make proposals for public policy. It argues that a quasi-public sector is needed to provide quasi-public goods such as export processing zones. Quasi-public goods are goods that have some public characteristics (nonexcludability, nonrivalry, and/or indivisility) but not all of them. In the case of the Mexican Silicon Valley, CANIETI and CADELEC could be qualified as quasi-public actors providing coordination initiatives among private enterprises.

The flowchart approach stresses not only the quasi-public sector, but also one or more anchor firms that have increasing returns to scale. Eventually the industrial landscape evolves into a complex of the anchor firm and related firms, together creating economies of scale. As a consequence, an industrial cluster takes shape around this complex, giving a boost to regional economic growth. The policy recommendation made by the flowchart approach is a sequence of measures, beginning with the establishment of an industrial zone, followed by inviting an anchor company, and then by promoting this firm's related companies for investment in the industrial zone. The sequence ends with reducing government's role to induce further competition.

The underlying model for this approach is a one-sector model, which could implicitly conceal the harmful effects of preferential treatment of one sector over other sectors, highlighted in two-sector models (Hamada (1974) in the case of duty-free zones, Brecher and Diaz-Alejandro (1977) in the case of foreign direct investment, and Dixit and Grossman (1986) in the case of strategic trade policy). Also, it is difficult to pick winners *ex-ante* (sector-wise and/or company-wise). Finally, global commodity chain literature highlights that, within each industry, an anchor firm could be different for each supply chain. In the case of the textile industry, the final distribution company is an anchor firm, while in the automobile industry the assembling company plays the anchor role. Thus, it is difficult to correctly anticipate all these elements before designing a quasi-public sector for establishing an export-processing zone.

In spite of all these reservations above, quasi-public efforts certainly involve many trial-and-error experiential learning processes. Therefore, from a heuristic point of view, quasi-public efforts such as export-processing zones could be useful in encouraging learning activities, which could contribute to human capital enhancement. In the case of the Mexican Silicon Valley, CADELEC could be a success story, while promotion of the software industry is still a long way off. However, through sweat and tears, certain local technological and management bases have put down roots, culminating in the establishment of a few local-capital companies.

6.2. The flowchart approach in the case of the Mexican silicon valley

6.2.1. Who is the quasi-public actor?

Palacios (2005) shows that industrial parks were mostly established in this cluster in the late 1990s, and indicates chronologically that they were not essential for the start of the cluster or its early development. Thus, a quasi-public good in the form of industrial parks is not necessarily the conduit of the growth of this cluster. In contrast, CANIETI and CADELEC could be categorized as important quasi-public actors. The caveat is that the members of CANIETI are local professional managers working in multinational companies that are competing globally with one another. This could have both positive and negative implications for cluster promotion. Positive effects could include avoidance of "white elephants": local managers are the ones who know the market best. Negative effects could result if cluster promotion becomes the managers' trivial side job. These managers are remunerated by their companies, not by the business association, meaning that they may easily dismiss factors external to the cluster.

6.2.2. Who is the anchor firm?

The main feature of the Mexican Silicon Valley is manufacturing services provided by CEMs; the electronics industry in the cluster does not exhibit the implicit long-term relationships typically found among, for example, Japanese companies. Certainly, as we have argued, historical factors such as IBM's establishment during the import-substitution industrialization period favored the emergence of an electronics cluster in the Guadalajara area. However, this does not necessarily imply that one BC (IBM or HP) or one CEM among many acts like an anchor firm in Guadalajara. It could be said that all the BCs and CEMs have formed one anchor complex, which in turn attracts specialized suppliers. This anchor complex was established due to the locational advantages described in section 3.2.

6.2.3. Who are the related firms?

Certainly, specialized suppliers and design centers could be categorized as related firms attracted to the anchor complex in the Guadalajara area. With related firms included in the mix, the Mexican Silicon Valley forms an industrial cluster. It is important to note that CANIETI and CADELEC are heavily involved in cluster promotion efforts. CANIETI and CADELEC are also dedicated to technology cooperation between local universities and companies.

7. Conclusion

This chapter examined how and why Guadalajara has risen to prominence as "the Mexican Silicon Valley." First, we reviewed the export and investment performance of the cluster. The very impressive growth in the last half of the 1990s was contrasted with stagnant performance in the 2000s. We also explained the basic structure of the cluster and the principal enterprises in the cluster. This cluster is largely made up of Brand Companies (BCs), Contract Electronic Manufactures (CEMs), specialized suppliers, and design centers. The enterprises are largely foreign companies, but there are some domestic companies.

Next, we scrutinized why and how major companies chose the Guadalajara area as a production site. There were various factors, such as macro-economic contingency (massive exchange rate devaluation in 1994), market proximity to the U.S., a relatively stable labor market environment, skilled human resources trained by universities, history (prior investment), and institutional support. We also highlighted those who contributed to combining the above-mentioned locational advantages – local professional managers. These individuals promoted the cluster through the local business association.

We explored the challenge of the cluster from the late 1990s onward. With the end of the IT bubble in the U.S. and the rise of China, the cluster has been faced with formidable challenges. Many companies are trying to produce higher value-added products in response to these challenges. Meanwhile, the cluster has kept muddling through in the global market. Local linkages are in the process of formation, largely in the area of human resources. Local companies are emerging, but these are still few in number. Financial linkages are still weak because multinationals can tap U.S. capital markets.

Finally, we tried to look at "the Mexican Silicon Valley" through the lens of the flowchart approach to industrial cluster policy. We claim that the local business association (CANIETI) and its spin-off organization (CADELEC) could be considered as quasi-public actors. They provide the cluster with various coordination and information services that can be interpreted as provision of quasi-public goods. These quasi-public actors are important in the sense that they carry out important processes of trial-and-error. Through these collective learning experiences, firms in the Mexican Silicon Valley can grow further in the competitive global electronics market.

Acknowledgments

I conducted the interviews for this study in September 2005 in Guadalajara. I would like to thank Dr. Juan José Palacios and Mr. Jacobo González for our fruitful discussions. I would also like to thank the seminar participants at IDE-JETRO for their comments and discussions. Finally, I am grateful to Mr. Hiroyuki Ukeda and Mr. Yasushi Ueki for their extensive comments. The usual caveat applies.

Notes

1. The very low presence of Japanese companies is in remarkable contrast to the proliferation of Japanese companies in TV production along the Mexico-U.S. border. See Koido (2003) and ITAM (2005) on TV production clusters.

2. See Palacios (2004) for a more detailed description of the evolution of the cluster.
3. *Electronics Manufacturing* magazine (Sep. 1, 2004) published the list of top 100 contract manufacturers, which named Flextronics, Solectron, Sanmina-SCI, and Jabil Circuit as the No.1, No.2, No.3, and No.5 CEMs, respectively.
4. See Sturgeon (2002) on its industrial organization impact.
5. Interview conducted in the summer of 2001 with a CEM manger.
6. The Maquiladora regime as export promotion ended in the end of 1999 as part of the transition framework of NAFTA.
7. See Dedrick, Kraemer, and Palacios (2001) on impacts of liberalization on Mexico's computer sector.
8. Of course, the other is Canada.
9. According to CADELEC (2005), since 2000 the index of rotation of workers has been less than 5 percent, which means that less than 5 percent of effective workers leave the workplace during the period.
10. See Rivera Vargas (2002) on technology transfer between universities and industry.
11. According to ITAM (2005), the ratio of direct workers relative to indirect workers in multinational electronics companies in Jalisco state is 2:7, while that of Baja California state is 4:6. Direct workers are those who have direct contact with the product, while indirect workers are engineers and administrative staff.
12. In an interview on September 2005, a CADELEC officer stated that there are more than 10 universities in the area: Universidad de Guadalajara, Universidad Autónoma de Guadalajara, ITESO, Tec de Monterrey, UNIVA, Panamericana, Tec Milenio, U del Valle de México, Unitec, Universidad Cuahtémoc, and La Salle.
13. Motorola started local production in 1968, and IBM started producing electronic typewriters locally in 1975. See Borja (1995) regarding the industrial promotion regime in the electronics sector up to the 1980s.
14. La Porta et al. (1998) show that in developing countries ownership is more concentrated than in developed countries.
15. Many of these were born and bred locally.
16. See Thacker (2000) for more on Mexican domestic big business and business associations in the early 1990s.
17. In simple terms, when one company fails, they seek to be hired by other companies.
18. A traditional Japanese electronics firm, in contrast, is engaged in implicit long-term contracts with exclusive suppliers.
19. CADELEC (2005) is one example.
20. Dussel Peters (2003) explains the current export promotion regime.
21. According to Palacios (2005), On Semiconductor in 2002 closed down the plant it had acquired from Motorola in 1999, sold it to a local concern, and moved its local operations to its plants abroad.
22. Our interview confirms this is the strategy of IBM and Intel. See Luhnow (2004) for other cases.
23. See Rivera Vargas and Regino Madlonado (2004) for more information.
24. According to Conacyt (2002), 12 establishments in the office machine and computer industry and 18 establishments in the electric machinery sector in Jalisco state received ISO 9000 from 1991 through 2002.

Bibliography

Borja, Arturo. *El Estado y el Desarrollo Industrial: La Política Mexicana de Cómputo en una Perspectivea Comparada*. México: CIDE, 1995.

Brecher, Richard A., and Carlos F. Díaz-Alejandro. "Tariffs, Foreign Capital and Immiserizing Growth." *Journal of International Economics* 1977: 7, no. 4, pp. 317–22.

Cadena Productiva de la Electrónica, A.C. (CADELEC). *Estudio de Rotación de Personal: Cluster de la Industria Electrómica de Jalisco.* México: CADELEC, 2005.

Consejo Nacional de Ciencia y Tecnología (CONACYT). *Estudio sobre los Establecimientos Certificados en ISO-9000 en México, 2002.* México: CONACYT, 2003.

Dedrick, Jason, Kenneth L. Kraemer, and Juan J. Palacios. "Impacts of Liberalization and Economic Intergration on Mexico's Computer Sector." *The Information Society* 2001: 17, no. 2, pp. 119–132.

Dixit, Avinash, and Gene Grossman. "Targeted Export Promotion with Several Oligopolistic Industries." *Journal of International Economics* 1986: 21, no. 3/4, pp. 233–49.

Dussel Peters, Enrique. "Ser Maquila o no ser Maquila, Es ésa la Pregunta? (To Be or Not To Be Maquila. Is That the Questions? With English Summary)" *Comercio Exterior* 2003: 53, no. 4, pp. 328–36.

Hamada, Koichi. "An Economic Analysis of the Duty-Free Zone." *Journal of International Economics* 1974: 4, no. 3, pp. 225–41.

Instituto Nacional de Estadística Geografía e Informática (INEGI). *Industria Maquiladora de Exportación, Noviembre 2005.* México: INEGI, 2005.

Instituto Tecnológico Autónomo de México (ITAM). *La Industria Electrónica en México: Diagnóstico, Prospectivea y Estrategia*. México: Centro de Estudios de Competitividad, ITAM, 2005.

Koido, Akihiro. "La Industria de Televisores a Color en la Frontera de México con Esutados Unidos: Potencial y Límites del Desarrollo Local." *Comerico Exterior* 2003: 53, no. 4, pp. 356–72.

La Porta, Rafael, Florencio Lopez-de-Silane, Andrei Shleifer and Robert W. Vishny. "Law and Finance." *Journal of Political Economy* 1998: 106, no. 6, pp. 1113–1155.

Luhnow, David. "As Jobs Move East, Plants in Mexico Retool to Compete." *Wall Street Journal*. March 5, 2004, p. 1.

Palacios, Juan J. "El Valle del Silicio Mexicano: Orígenes, Evolución y Características del Complejo Industrial de la Electrónica en Guadalajara." *Condiciones y Retos de la Electrónica en México.* Eds. Enrique Dussel Peters and Juan José Palacios Lara. México: NYCE, 2004.

Palacios, Juan J. "Economic Agglomeration and Industrial Clustering in Developing Countries: The Case of the Mexican Silicon Valley." *Research Report on the project Comparison of Industrial Agglomerations between Asia and other Regions prepared for the Institute of Developing Economies, Japan External Trade Organization (IDE-JETRO)* 2005.

Rivera Vargas, María Isabel. *Technology Transfer via University-Industry Relationship: The Case of the Foreign High Technology Electronics Industry in Mexico's Silicon Valley.* New York: RoutledgeFalmer, 2002.

Rivera Vargas, María Isabel, and Juan Regino Maldonado. "Aprendizaje Tecnológico en los Proveedores de la Industria Electrónica, Guadalajara, México." *Comercio Exterior* 2004: 54, no. 3, pp. 196–206.

Sturgeon, Timothy J. "Modular Production Networks: A New American Model of Industrial Organization." *Industrial and Corporate Change* 2002: 11, no. 3, pp. 451–96.

Thacker, Strom C. *Big Business, the State and Free Trade: Constructing Coalitions in Mexico.* Cambridge: Cambridge University Press, 2000.

9 Conclusion

Masatsugu Tsuji and Akifumi Kuchiki

For the last several years, we have studied industrial agglomeration and made an attempt to explain why firms agglomerate (see for example, Kagami and Tsuji (2003), Kuchiki and Tsuji (2005)). Industrial agglomeration is not a single phenomenon, but it has lots of variety: Kagami and Tsuji (2003) summarize agglomerations into five categories: (1) clusters where locally specialized items are produced; (2) clusters where a large core (anchor) firm has many subcontractors; (3) urban processing clusters where many basic production processes are treated; (4) clusters where advanced technologies and their services are transacted – such as information and communication technology (ICT) clusters; and (5) government-led industrial parks and estates often seen in developing countries. In order to explain these agglomerations, the theoretical frameworks of agglomeration from Marshal to Kurgman, Porter, Fujita and Thisse are available. Economies of scales and transportation costs are key issues in their analyses.

In addition to theoretical issues, agglomeration tends to relate more to realistic interests. East Asian countries have been successfully attracting foreign firms into their regions and these agglomerations become an engine of their economic growth. The same examples are easily found in other areas such as India, Mexico, and Eastern Europe. In accordance with globalization of the world economy, funds, human resources, information, and technologies are crossing national borders. Firms are even developing globally their business activities such as production, marketing, and R&D so as to capture better managerial resources or markets. Not only developing but also developed economies are competing with each other to attract all kinds of firms, domestic or international. Whether they are successful is essential to their future economic development. In this sense, agglomeration is a key issue to policy authorities of central as well as local governments.

Based on these backgrounds, this volume is an outcome of the research we have undertaken to construct a prototype model which formulates the essence of current agglomeration in East Asia. The model presented here is referred to as "the flowchart approach". The flowchart approach can explain why firms agglomerate in East Asia, and therefore it can be applied in the construction of an industrial cluster policy. In this sense, the flowchart approach presents a growth strategy for regional as well as national economies by prioritizing factors in order to promote industrial clusters: that is, the flowchart approach tries to answer such questions as what conditions are required to form a new agglomeration and to develop it to an industrial cluster.

Part I of this volume propounds the theoretical foundations of the flowchart approach, while Part II presents case studies in different countries which show it's effectiveness. All chapters examine and verify the flowchart approach. Some of the points discussed in different chapters are as follows. Examination of the information technology industry in Austin, Texas, finds that (i) living conditions such as low living costs and various cultural activities are important in attracting excellent human resources, while (ii) universities are also a part of the infrastructure which supplies good human resources to the region. Cases of industrial clusters in India show that the geographical location of ports is crucial. Regarding anchor firms, scrutiny of the automobile industry cluster in Guangzhou, China, found that Honda, Nissan and Toyota were anchor firms, and that inviting them to the region have been success factors in the industrial cluster policy, as realized by the mayor of Guangzhou. This shows that local government leadership is also a crucial factor in the success of the industrial cluster policy.

Another characteristic of this volume is the empirical study of such countries as Japan and Thailand. These two case studies examine empirically the verifiability of the flowchart approach. The study of Thailand proves rigorously that anchor firms came to the region first, and related firms followed; this is one of the important hypotheses in the flowchart approach. In Japan, building infrastructure – such as incubation facilities, system supports, and human resources training – is conducted as industrial cluster policy. Without suitable planning, it is difficult for those cluster policies to be successful.

As stated above, this volume focuses on the flowchart approach to explain industrial agglomeration, but it should be noted that the approach cannot explain all phenomena. Further research is required to identify for what kind of industry or at what the level of industrial development the flowchart approach can explain agglomeration best. Moreover, some East Asian economies have been achieving the stage of innovation, rather than that of a simple production base. In those areas, endogenous innovation or R&D activities are occurring. These are natural outcomes of agglomeration, as emphasized by Fujita and Thisse (2002). Agglomerations create further exchange of information, know-how, and even tacit knowledge, and these promote the endogenous flow of innovation or R&D. We have to modify the flowchart approach to explain these processes.

Another issue is related to international comparative studies. The stage of economic development is different between economies; some are more developed and some are not. The level of agglomeration is also different. Through the flowchart approach, indexes of agglomeration can be constructed. These indexes will be helpful in formulating more suitable policy measures to promote agglomeration in a region.

References

Fujita, Masahisa and Jacques-François Thisse (2002), *Economics of Agglomeration: Cities, Industrial Location, and Regional Growth*, Cambridge: Cambridge University Press.

Kuchiki, Akifumi and Masatsugu Tsuji (eds) (2005), *Industrial Clusters in Asia: Analysis of their Competition and Cooperation*, London: Palgrave Macmillan.

Tsuji, Masatsugu, Emanuele Giovanetti and Mitsuhiro Kagami (eds) (2007), *Industrial Agglomeration and New Technologies: A Global Perspective*. Cheltenham: Edward Elgar.

Appendix: Theory of a Flowchart Approach to Industrial Cluster Policy

Akifumi Kuchiki

1. Introduction

Policies for forming industrial clusters, or industrial cluster policies, play an important role in regional development in many countries, including countries in the European Union and Asia. A cluster is a geographic concentration within a nation or region of interconnected companies, specialized suppliers, service providers, and associated institutions in a particular field (a good example is Western Scotland). East Asia contains a number of examples of practical application of industrial cluster models and policy measures. Japan's Ministry of Economy, Trade, and Industry (METI) has approved 19 industrial cluster plan (see Mitsui, 2003). Malaysia's Second Industrial Master Plan of 1996–2005 includes policies to promote the development of competitive clusters in the electronics industry. The Industrial Estate Authority of Thailand (IEAT), established in 1972, is a state enterprise under the jurisdiction of the Ministry of Industry that is responsible for carrying out the country's industrial development policy. Thailand's National Economics and Social Development Board (NESDB) plans, along with the Industrial Estate Authority of Thailand (IEAT), to implement regional development policy by promoting the formation of industrial clusters. In June 2004, the NESDB announced that within three years, it intends to form eight clusters as the core of the automobile and electronics industries in four Thai regions. The IEAT plans to establish industrial zones specifically for industrial clusters, such as the Eastern Sea Board Industrial Zone's automotive industry cluster.

Porter (1990) recommended what he termed a "diamond approach" to cluster policy. The four points of the diamond represent factor conditions, demand conditions, related industries, and firm strategy/rivalry. These are the four basic attributes that affect regional productivity and innovation.

Each of these four attributes is self-reinforcing over time. Each has a unique and important role to play in a region's business environment, and all four operate together as a system. However, the diamond approach is not a policy approach in the true sense of the word, because the four factors do not clarify the role of government.

Markusen (1996) classified industrial agglomerations into four types, but these types do not explain the common patterns of Asian regional development. Kuchiki (2003) found that conditions for forming new clusters in northern Vietnam included: (a) industrial zones, (b) capacity building of physical infrastructure along with institutional reforms regarding investment procedures, (c) anchor firms in the manufacturing industry, and (d) anchor firms' related firms. That study showed that industrial zones, together with a combination of infrastructure and institutions, played crucial roles in the development of industrial agglomerations. However, no one has yet formulated a theory of the sufficient conditions for the success of industrial cluster policy.

In this appendix, we develop a "flowchart approach to industrial cluster policy," by proposing sufficient conditions for the formation of industrial clusters of the type often found in the manufacturing industry in Asia, and by deriving the sufficiency of these conditions for enhancement of regional economic growth. We show that the typical pattern of industrial cluster formation in East Asia includes what we call "quasi-public goods." We then demonstrate that industrial cluster policy enhances economic growth under a production function of "increasing returns to scale," and we show that critical amounts of production necessary for "scale economies" which must be in place for firms to invest in clusters can be determined. The concepts of (1) "quasi-public goods," (2) "increasing returns to scale" and (3) "scale economies" are crucial to this theoretical framework.

(1) Industrial zones are provided as quasi-public goods, both by organizations in the quasi-public sector and by firms in the private sector. (2 Industrial cluster policy can enhance regional economic growth in cases where an anchor firm operates under a production function of "increasing returns to scale." Markets in Asia are typically large enough for anchor firms to attain "increasing returns to scale." (3) The fixed capital of companies related to the anchor company determines the minimum optimal size of production necessary for "scale economies."

It is noted that we derive not necessary conditions but sufficient conditions in order that the industrial cluster policy of our flowchart approach functions well to form an industrial cluster. We focus economic agents in the flowchart, find their sufficient conditions and show that a step in a flowchart will move forward to the next step if each economic agent at each step satisfies its sufficient conditions. This appendix shows the following two sufficient conditions at two steps of the flowchart.

First, a public corporation in the quasi-public sector should provide quasi-public goods, and firms in the private sector can provide the same quasi-public goods since they can efficiently provide "optional" quasi-public goods. That is, economic agents in both the private sector and public sectors can efficiently provide industrial zones since the industrial zones belongs to the optional quasi-public goods. For example, at first the Industrial Estate Authority of Thailand in the quasi-public sector established export processing zones in the early 1980s. Then private firms such as Japanese trading firms established export-processing zones in Thailand in the later half of the 1980s.

Second, we illustrate a sufficient condition for an anchor firm to grow in a region by industrial cluster policy. The sufficient condition is that a production function of an anchor firm is attaining "increasing returns to scale."

Section 2 defines "quasi-public goods," and applies this definition to industrial zones and capacity building. Section 3 demonstrates that industrial cluster policy in East Asia can enhance regional economic growth in cases where anchor firms operate under a production function of "increasing returns to scale." Section 4 explains how markets in China are large enough for firms related to anchor firms to attain "scale economies." Section 5 discusses these results and concludes the paper.

2. Quasi-public goods

2.1. The Role of the quasi-public goods of industrial zones in indus trial cluster policy

This section includes a definition of quasi-public goods and clarifies the role of local government in implementing industrial cluster policy. Both the private and quasi-public sectors can optimally provide quasi-public goods as part of industrial cluster policy.

Samuelson (1954) made a bi-polar distinction between pure public and pure private goods in the real world. Between these extremes fall quasi-public (non-pure) goods, which seem to have characteristics of both pure public and pure private goods. Samuelson (1955) conceded that many goods commonly defined as public goods do not fit his definition, and many articles have been published that allow further blurring of the distinction between pure private and pure public goods.

This section attempts to clarify the confusion that has arisen in discussing public goods, by revealing the assumptions implicit in Samuelson's model. For that purpose, we define three criteria that characterize goods. A taxonomy of these articles may be developed by utilizing these criteria.

Emphasis will be placed on the criteria of *excludability/non-excludability*, *rivalness/non-rivalness*, and *optionality/non-optionality*. Each characteristic can be defined mathematically, and a model can be built that describes both public goods and quasi-public goods as well. Arguments for optimal conditions of goods in terms of the three characteristics may then be made.

Some propositions and implications will then be put forth. Even if a good has rivalness, market failures may still occur if that good is non-excludable and non-optional. Vertically added demand curves may also be considered. Even if a good is non-excludable and non-rival, the individual marginal rates of substitution need not be added vertically for that good to be optional. Our analysis may also be a clue to solving the popular dispute between Samuelson (1964) and Minasian (1964). An important point is that the property of non-optionality is crucial in optimality conditions.

"Quasi-public optional goods" may be defined as non-excludable or non-rival, and as optional. From the point of inequality, it may be concluded that quasi-public optional goods should be decentralized according to peoples' preferences. It is often said that a mixed economy consists of the public and private sectors. However, it seems that the importance of the quasi-public sector, which belongs to neither the public nor private sector, has recently increased. The concept of quasi-public optional goods is useful in clarifying the role of the quasi-public sector. This section illustrates the most desirable quasi-public goods that can be supplied by both the quasi-public and the private sector.

Section 2.2 presents four criteria for classifying goods. In section 2.3, a model is built that includes some kinds of quasi-public goods. Section 2.4 contains optimality conditions. Finally, section 2.5 shows that industrial zones with tax incentives and one-stop services have the properties of quasi-public goods.

2.2. Criteria for classifying goods

This section includes definitions of non-excludability, non-rivalness, non-optionality, and non-indivisibility.

2.2.1. [NE] Non-excludability, [E] Excludability

A non-excludable good is one in which the supplier is not free to exclude individuals from using it at small or zero costs once the good is produced. Suppose that x units of a good (X) are produced, and that the ratio of these available to an individual (i) is k_i ($0 \leqq k_i \leqq 1$). Then the level of X available to the individual is xk_i. A non-excludable good is one in which k_i is determined at the same time that the good X is produced. It is given to the supplier, and the supplier cannot change k_i. An excludable good is one in which k_i can be changed by the supplier at nearly zero cost if that supplier wishes. Thus, for excludable goods, k_i is variable and will in general take different values for different suppliers.

2.2.2. [NO] Non-optionality, [O] Optionality

A non-optional good is one in which a demander i cannot change consumption level (x_i) freely once an available consumption level is determined as xk_i.

A constraint on the demander i is that the demander must consume all available levels of x in the case of a non-optional good. The available level is xk_i. Thus, $x_i = xk_i$. In the case of an optional good, the maximum available level is also xk_i, but demanders can change levels of consumption. The demander can choose a value of k_i from zero to one ($0 \leqq k_i \leqq 1$).

Dorfman (1969), as quoted by James (1969), indicates that "there are certain goods that have the peculiarity that they are available to everyone, and no one can be precluded from enjoying them whether he contributed to their provision or not." In other words, a non-optional good is one for which some positive consumption level is exogenously imposed, and any attempt to deviate, either upward or downward, from this level requires additional expenditures.

2.2.3. [NR] Non-rivalness, [R] Rivalness

A non-rival good is one in which "all enjoy in common in the sense that each individual's consumption of such a good leads to no reduction from any other individual's consumption of that good" (Samuelson, 1954). In general, non-rivalness does not mean that the same precise product quality is available to each demander. For example, demander 1 who lives close to the police station is considered to have better protection than the demander 2 who lives far away from it. That is,

$$k_1 > k_2$$

It may therefore be deduced that in the case of a rival good, the following must hold:

$$k_r = \{k_i \mid \Sigma\, k_i \leq 1, \qquad 0 \leq k_i \leq 1\}$$

In the case of a non-rival good, on the other hand, the condition is:

$$k_n = \{k_i \mid \Sigma\, k_i > 1 \quad \text{or } \Sigma\, k_i \leq 1,\ 0 \leq k_i \leq 1\}$$

2.2.4. [ID] Indivisibility, [D] Divisibility

Divisible goods come in discrete units, some of them quite large or "lumpy." Usually, an indivisible commodity is defined as an "integer" (see Gomory and Baumo, 1960; Frank, 1969; and others).

Figure A.1 depicts the process from production to consumption, and explains which process relates to each characteristic. Non-excludability or excludability relate to the supply process, and non-optionality or optionality to the demand process. We may analyze articles dealing with theories of quasi-public goods, according to which of these characteristics the articles assume such goods to have. For example, Holtermann (1972) assumes [D] [NE] [R] [O] [ID], and Davis and Whinston (1967) assumes [ID] [NE] [R] [O]. Samuelson's pure collective goods correspond to [NE], [NO], and [NR], k_i=1 for all i. Theoretical studies of indivisible goods have been published with

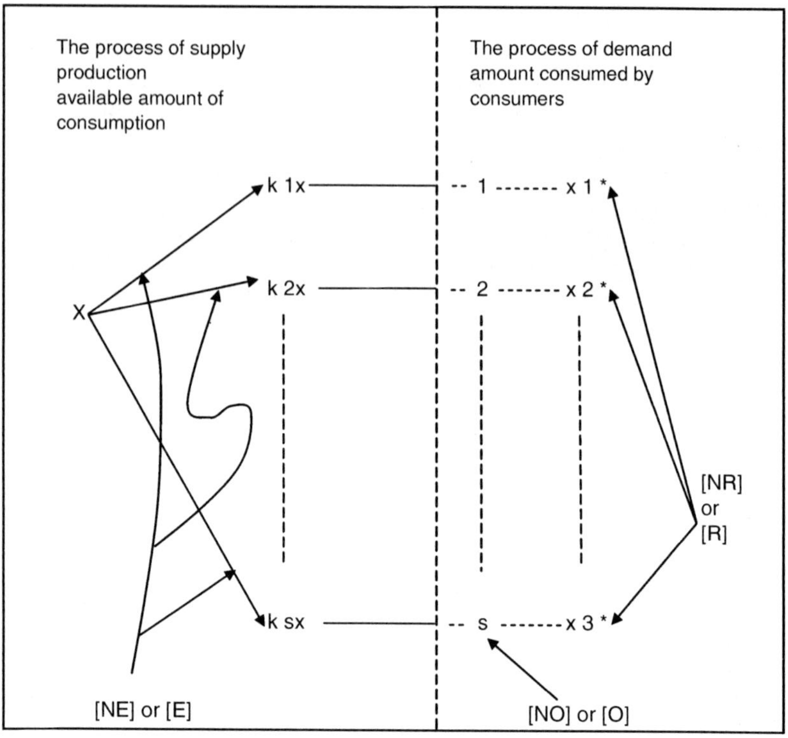

Figure A.1 Definiton of goods
Source: author.

increasing frequency, because these tend to be more relevant for solving real world problems. It is necessary to analyze the cases that remain untouched. Hereafter, attention will be given to [NE] or [E], [NR] or [R], and [NO] or [O] (see Appendices 1 and 2).

2.3. The model

Consider the following cases and examples corresponding to each:

1. V: [NE] [NR] [NO]:	national defense, national security (food self-sufficiency)
2. W: [NE] [NR] [O]:	television broadcasting, radio frequencies
3. X: [NE] [R] [O]:	a limited supply of oxygen
4. Y: [NE] [R] [O]:	outdoor circuses, green utilities (agriculture)
5. T: [NE][NR][O]:	research
6. Z: [E] [R] [O] :	bread

Proposition 1: If a good is excludable, each k_i is variable for the demander i. Each demander must determine the optimal value of k_i; otherwise, k_i remains indeterminate, and therefore cannot be non-optional.

The economy has six kinds of goods, including two kinds of pure public and private goods (V,Z), and four kinds of non-pure goods (W,X,Y,T). Here, $(V_1,\ldots,V_j; W_1,\ldots,W_k; X_1,\ldots,X_\ell; Y_1,\ldots,Y_m; Z_1,\ldots,Z_n; T_1,\ldots,T_p)=(V,W,X,Y,Z,T)$. In the following equations, superscripts refer to persons and subscripts to goods.

When properties of goods are considered, the constraints of these goods are as follows:

$$V_a^i = v_a^i V_a \tag{1}$$

where $0 \leq v_a^i \leq 1$, v_a^i are constants, and $\sum_i v_a^i > 1$ is possible

for all i, a=1, …, j.

$$W_b^i \leq w_b^i W_b, \tag{2}$$

where w_b^i correspond to v_a^i in (1).

$$\sum_i Z_e^i \leq Z_e, \ e=1, \ldots, n \tag{3}$$

Constraint (2) indicates that no more goods can be consumed than is allowed by the quantity of goods available to the individual i. Here v_a^i, w_b^i, x_c^i and y_d^i corresponding to k_i are used in section 2.2.

Proposition 2: The optimality condition for T, in which non-rivalness plays a crucial role, is the same as that for W. In the case of T, however, it is always necessary that t_f^i the corresponding to k_i be equal to unity for all i (see Appendix 3).

Proposition 3: Both rival and non-rival goods have essentially the same optimality condition between W and Y. The result is similar to the relationship between V and X (see Appendix 3).

It may be assumed that the production possibilities for the economy are described by a well-behaved transformation locus:
$F(V, W, Z) \leq 0$.

Here V represents pure public goods, W represents non-excludable, non-rival and optional goods, and Z refers to pure private goods. Hereafter, T is omitted because of Proposition 2. Attention is also placed on V and W because of Proposition 3.

Each individual's utility function (u^i) and the social welfare function (U) may be assumed to be the same as those of Samuelson (1964). The maximization problem is given in Appendix 4.

2.4. Optimality conditions

The results of V and Z are well known. That is, if pure private goods Z are chosen, then an individual should consume an amount that equals his weighted marginal utility to the price of the good. The multipliers can be interpreted as shadow prices. For each pure public good V, the sum of the marginal rate is equal to the inverse of that private good in the consumption of the pure one. The distinction between public goods and private goods has been proposed as being the non-rivalness of public goods (Musgrave, 1969). However, the crucial characteristic in making this distinction is not that of non-rivalness, but rather those of non-optionality and non-excludability (Appendix 4). Musgrave (1969) says: "Due to the non-rivalness of consumption, individual demand curves are added vertically rather than horizontally as in the case of private goods."

This statement is not always true in terms of the definitions given in this article.

Proposition 4: If a good is optional, then even if it is non-excludable and non-rival, the optimality condition does not require

$$\sum MRS^i = MRT,$$

where MRS and MRT denote the marginal rate of substitution and the marginal rate of transformation, respectively.

Consider the extreme,

$$MRS^i = MRT, \text{ for } i \in M_b.$$

It is possible that the "benefit principle" holds in the case of optional goods W. The method of charging tolls probably succeeds from the viewpoint of optimal allocation of resources, for users cannot choose in ways that do not reveal their true preferences. A similar argument also applies to non-excludable, rival, and optional goods Y (see Appendix 5).

It is possible to suppose that such situations will occur. In general, however, the sets M_b will consist of more than one person. Attention may be placed on the sets M_b alone. Here it is very important that, when obtaining optimality conditions, members of the sets N_b can be neglected. In general, these people do not want to consume goods W.

This proposition has some implications. First, Samuelson (1955) argues: "A point on the efficiency frontier requires equality between the vertically added marginal rates of substitution of all men for the public and private goods."

However, it is possible that, in the case of W([NE][NR][O]) or Y([NE][R][O]), different individuals' MRS need not be added vertically. It is thus possible to exclude from consideration those who do not belong to the sets M_b.

2.5. Industrial zones as quasi-public goods

Consider an export-processing zone with one-stop services and tax reductions with exemptions to promote exports, as shown in Table A.1. The zone, representing the quasi-public goods of X, is set up as a single unit on a 300-hectare plot, and includes a town equipped with infrastructure. The zone's

Table A.1 Export processing zones as quasi-public goods

	NR or R	NE or E	NO or O	
I. Industrial zone (200ha)	R	NE	O	X
II. Infrastructure				
(1) Electricity	R	E	O	Z
(2) Roads	NR	NE	O	W
(3) Water supply	R	E	O	Z
III. Instititions				
(1) Preferences	NR	E	O	
(2) One- Stop Services	NR	E	O	
	Rivalness	Excludable	Optional	

Note: Nr, R, IND, D, NE, E, NO, and O denote non-rival, rival, non-excludable, excludable, non-optional, and optional, respectively.

Source: author.

town, consisting of factories, housing, and amusement facilities, is indivisible, but it is also excludable and optional to each company. An office at the zone provides services and procedures for companies to establish their plants. One-stop investment services are rival and optional. Taxes such as income and import tariffs of companies are reduced or exempted. Tax incentives in an export-processing zone are non-rival and non-optional. In the next section, G denotes the total government investment in an export-processing zone.

The Asian experience of economic growth exhibits certain characteristics that are common across many countries. The economic sector is separated into the private sector and the public sector. Further, goods are categorized into those subject to "market competition" and those that generate "market failures." The private sector bears the responsibility for market competition (Arrow D in Figure A.2), and the public sector bears the responsibility for market failures (Arrow B in Figure A.2). Intervention in the private sector by the public sector is the active industrial policy of Arrow C in Figure A.2. This is a textbook approach in economics.

Figure A.2 shows the role of the public sector in market failures, which has been thoroughly analyzed in the past. Experiences in Asia, however, show that private-sector corporations can cover market failure losses, as Arrow A in Figure A.2. Arrow A represents a new activity that has resulted from globalization. For example, multinational corporations (IPP's or Independent Power Producers) supply electrical power to Asian countries.

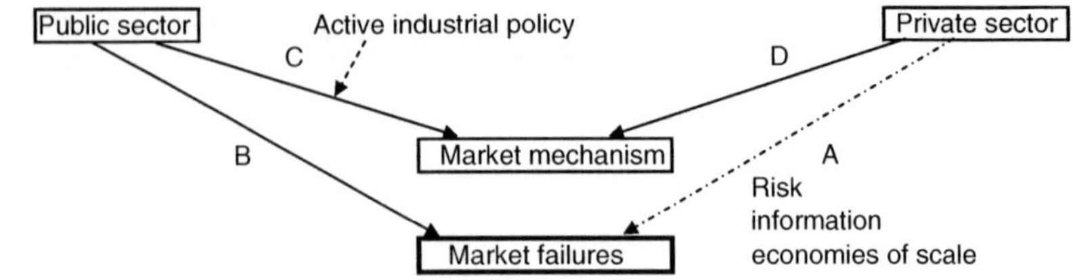

Figure A.2 The role of the private sector in market failures (Importance of economic agents)

Source: author.

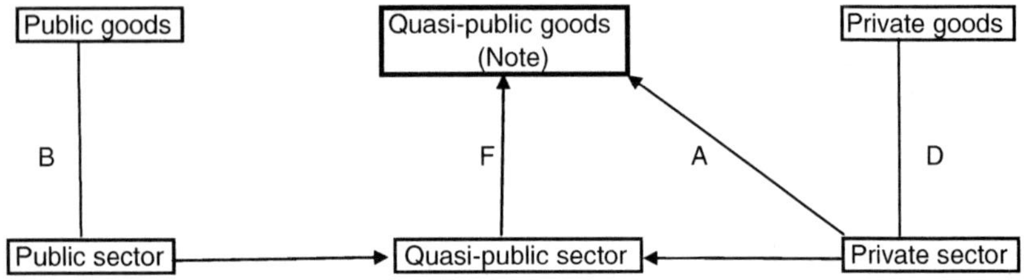

Figure A.3 Importance of quasi-public goods

Note: These include excludable vs. nonexcludable, rival vs. nonrival, and divisible vs. nondivisible.

Source: author.

BOT (Build – Operate – Transfer) is a system by which the private sector supplies infrastructure. Certain infrastructure is categorized as a public good, and thus represents one type of market failure. Japanese trading corporations supply infrastructure in Asia using this system.

Figure A.3 clarifies the role played by the quasi-public sector in economic growth. This sector is positioned midway between the private and public sectors. Arrow F in Figure A.3 shows the role of the quasi-public sector in providing quasi-public goods. As an example, the Industrial Estate Authority of Thailand played an important quasi-public sector role during the construction of industrial zones in Thailand in the latter half of the 1980's; these industrial zones effectively introduced foreign investment to Thailand. The point that requires emphasis here is that quasi-public goods supplied by the quasi-public sector played a leading role in the early stages of economic development. Note that Arrow A shows the role of the private sector in providing quasi-public goods – multinational corporations can establish export-processing zones.

Another common characteristic of the Asian growth experience is that the quasi-public sector plays an important role in the process of economic development, as shown in Figure A.3. One typical example of a market failure is a pure public good. A pure public good is non-exclusive, non-competitive, and indivisible, as explained in Section 2.2. Numerous quasi-public goods, such as export-processing zones, exist. These may lack one or two of the properties of a public good, but not all three. The role of the quasi-public sector must be distinguished from that of the private sector by taking into account the properties of quasi-public goods in formulating an industrial cluster policy.

In some cases, it is desirable that quasi-public goods be supplied by economic agents in the quasi-public sector. So far, there has perhaps been too much discussion about the roles of the state or government, without sufficient analysis of the "quasi-public goods" that should be supplied by economic agents in the quasi-public sector.

3. Industrial cluster policy in East Asia

3.1. A regional growth model: increasing returns to scale in anchor firms

Asian experiences in the 1980's showed that industrial zones or export-processing zones (EPZs) in East Asia contributed to the generation of employment opportunities. EPZs limit job opportunities to people living near these zones. An interesting question is whether or not EPZs, as quasi-public goods, can enhance the aggregate growth of a region by forming industrial clusters.

In a pioneering work, Hamada (1974) analyzed EPZs. Grossman and Helpman (1991) built a model that takes into considerations both innovators and imitators of new technology. The purpose of this section is to determine conditions under which EPZs can enhance aggregate growth, and to examine whether or not reduction in tax rates for foreign investors is effective in enhancing aggregate growth. To that end, this section includes the application to EPZs of a model by Grossman and Helpman.

Based on experiences in East Asia, we can construct an EPZ model that explains a growth mechanism in East Asia in which industrial zones or export-processing zones link multinational companies who are pursuing cost reduction with governments of recipient countries who implement deregulation and preferential tax treatment. Specific focus is placed on the role of EPZs in East Asia. A behavioral theory relative to multinationals that invest in EPZs is also applied.

When a macro-economic growth model is constructed to analyze the effectiveness of EPZs in enhancing a nation's aggregate growth, it can be concluded that national income can be increased by reducing profit taxes if production functions of final goods of multinational companies located in EPZs have increasing returns to scale. Furthermore, governments should invite multinationals to EPZs if the costs of invitation are much cheaper than those of imitation, or if the number of intermediate goods of multinationals is large.

Section 3.2 includes the development of a growth model of an industrial cluster with industrial zones, and also includes analysis of a case where multinationals of country 1 are innovators and local firms of country 2 imitate innovations of the multinationals by paying the fee for imitation. Section 3.3 includes analysis of a case where, in addition to section 2, multinationals of country 1 make a decision about whether or not they will invest in EPZs of country 2. Conclusions are presented in Section 3.4.

3.2. The case of a country imitating innovations of multinational companies

In this section, a model of an industrial cluster with industrial zones is constructed, and it is shown that a region can achieve high rates of growth by adopting a preferential tax policy (t).

Two cases are considered:

- Case 1: Multinational companies of country 1 are innovators, and local firms of country 2 imitate the innovations of these multinationals at the cost of imitation,v.
- Case 2: In addition to the conditions of Case 1, multinational companies of country 1 make a decision about whether they will independently invest in EPZs in country 2.

In Case 1, it can be assumed that innovative products are intermediate inputs into the production of a single final good. Households can consume the final good output Y. The technology for producing the final output requires intermediate goods and the input of labor. Intermediate goods are produced by labor alone, and labor is also the sole input into research and development.

Two conclusions can be gleaned from this analysis (see Appendix 7):

- National income is increased by reducing the profit tax if the production functions of final goods of multinationals in EPZs of a developing country have increasing returns to scale.
- A government of a developing country should invite multinationals to its EPZs if the costs of invitation are much cheaper than the costs of imitation, or if the number of intermediate goods of multinational companies is large.

In the Hamada (1974) model, the welfare effects of an EPZ depend on the factor intensity of protected sectors in the domestic economy. Results in this article show that these effects depend on whether or not the multinational company's industry has increasing returns to scale. If this is the case, government plays a crucial role in inviting multinationals to EPZs and enhancing aggregate growth by reducing tax rates.

4. Sufficient condition for establishing a new plant in a cluster

4.1. Background and hypothesis

This section describes a case of increasing returns to scale with fixed capital. Suppose that firms use indivisible fixed capital. Fixed costs are costs needed at a very small level of production. Average costs decrease when fixed costs are large, and marginal variable costs are small. The average cost function may have a U-shaped curve when fixed costs exist; that is, average costs initially decrease and then increase. There exists an optimal size of production where the marginal costs are equal to the average costs.

An anchor company in the car assembly industry has a group of related companies that produce car parts, since a standard car is assembled from more than 25,000 components. This is true not only of Japanese companies, but also of the Korean company Hyundai and the American company General Motors. Clusters in the automobile industry exist in Guangzhou in southern China, Shanghai in central China, and Tianjin and Beijing in northern China. The distances from Shanghai to Tianjin and from Shanghai to Guangzhou are respectively 1200 and 2000 kilometers. Thus, component-producing companies must locate near anchor companies when anchor companies exceed certain levels of production. For example, the related companies of Toyota belong to a group called Kyohokai. The objective of Kyohokai is to produce cars to sell them to people living near these firms' plants.

In both Shanghai and Guangzhou, Japanese firms were interviewed on the subject of their optimal sizes. Three different types of component-producing company were identified, as shown in Tables A.2 and A.3:

1. Component-producing companies of Type I do not locate near their anchor companies, since they produce a variety of products for which their anchor company is not the sole market.
2. Component-producing companies of Type II belong to an anchor company's industrial group (such as a Japanese keiretsu, a term for a set of companies with interlocking business relationships and shareholders) and must locate their plant near that anchor company regardless of the production scale of those anchor companies.
3. Component-producing companies of Type III locate their plants near their anchor company only if they can attain minimal average costs at their minimal optimal size of production. Interview results showed that minimal optimal sizes ranged from 100,000 to 700,000 units.

In discussing whether or not the related firms of an anchor firm will invest in a cluster, the scale economies of the anchor firm must be taken into consideration.

Table A.2 Industrial cluster in Shanghai

Relationship with the anchor company	Name of the company	Product	Charecteristics
Type I	Company P	Audio for cars and mobile phones	Covers more than 90% of audio components in China
	Company D	Computerized electronic components	Necessary that, as a group company, it be located close to the anchor company with which it shares important data
Type II	Company T	Logistic company	Offers both the milk-run and cross-dock strategies
	Company K	Car lamps	Cluster process accelerates when the anchor company's sales of cars reach more than 300,000
Type III	Company A	Crank shafts for engines and cone rods	Cluster process accelerates when the anchor company's sales of cars reach more than 500,000
	Company S	Cold rolled steel sheets	Independent from the anchor company at a production size of more than 700,000

Source: author.

Hypothesis: The development of an industrial cluster depends on market demand, or the quantity of production of an anchor firm, and transportation costs. If an anchor firm expands its production size because of expected sales growth, then its related firms will establish new plants in the anchor firm's cluster.

4.2. Firms in Shanghai

The following examples illustrate how quantities of production are crucial to determining whether or not related firms will establish new plants.

4.2.1. Type I: company P

Company P, established in 1995 in China, produces four kinds of components, including speakers for car audio systems and mobile phones. In 2001,

Table A.3 Industrial cluster in Guangzhou

Relationship with the anchor company	Name of the company	Product	Charecteristics
Type I	Company T	Air-conditioners for autos and other	Does not specialize only in auto parts, and the plant is not required to be located close to the automobile production site
	Company HL	Key sets for cars	Depends on Honda Group
Type II	Company HA	Traiding company for Honda genuine auto parts	There are no economies of scale, and it is one of Honda's in-house departments
	Company M	Plastic Parts (e.g. Engine Covers)	Attains economies of scale, after production of 400,000 cars
Type III	Company FT	Clutches, brakes, and frames	Attains economies of scale, after production of 10,000 cars

Source: author.

Company P's research and development produced designs for die-casting processes for its products in Shanghai. Company P has the means to prevent these die-casting process designs from being stolen. It procures more than 90% of its components in China and sells about 8% of its products to Toyota in Tianjin, Honda in Guangzhou, and GM in Shanghai. Kamigumi, a Japanese logistics company, collects Company P's products for Toyota in what is commonly termed a "milk-run process," as it is similar to the process by which Japanese milk companies distribute milk bottles to homes in the morning. Company C is not inclined to outsource its components. It sells its car audio systems to car companies and sells its mobile phone speakers to Motorola in Tianjin. Thus, it has no plan to invest in Guangzhou in the near future.

4.2.2. Type II: company D

Company D produces computerized electronic components. As an electronic component producer, it has a close relationship with its anchor company. Company D shares important confidential data with the anchor company, including customer data. The company participates in the development of new car designs with its anchor company from the early stages of research and development. There are only a few large worldwide companies, such as Denso of Japan, Bosch of Germany, and Delphi of the US, that take part in this type of activity. These companies have no choice but to locate near their anchor companies.

4.2.3. Type II: company T

Company T is a logistics company that serves a Japanese automobile anchor company, adopting the aforementioned milk-run process as well as a cross-dock strategy. The milk-run process allows component-producing companies to prepare their products to be shipped at a time specified by their anchor company. Company T goes around to collect these products. The cross-dock strategy is a method by which the collected components are packaged as part of a "just-in-time" system, in order to provide components to the anchor assembly company efficiently. Company T has 5 milk-run routes; one point of collection provides enough components for 200 cars. Efficient logistics weaken incentives of component-producing companies in Shanghai to invest in Guangzhou and Tianjin. However, Company T must locate its anchor company where its components can be provided just in time. This is similar to the case of a *keiretsu* company. It is expected that sales of automobiles in China in 2010 will reach more than 10 million.

4.2.4. Type III: company K

Company K produces light bulbs for automobiles. Its transportation costs from Shanghai to Guangzhou represent more than 10% of total costs, since these products are bulky. The minimal optimal size of production is 300,000 units. The company cannot invest in Guangzhou if it is profitable for the company to transport products to Guangzhou, although the demand of Toyota in Guangzhou exceeds this firm's production by 300,000 units. This discrepancy is due to the fact that the company cannot convince its shareholders to establish a plant in Guangzhou if the plant will lose money.

4.2.5. Type III: company A

Company A produces crankshafts for engines and cone rods. As these components are used mainly for the central parts of engines, the material must be strong, but easy to process and cut. Company A is so capital-intensive that its investment totals ten million dollars. The optimal size of production of one line at a plant is 500,000 units. There are only three large Japanese companies in this industry. Company C has no plan to invest in Guangzhou, since it is located in Shanghai and transports its products all over China.

4.2.6. Type III: company S

Company S is a forging company. Cold rolled steel sheets used in the automobile industry are not produced in China, and production of this type of sheet requires the high technology of "deep-drawing." Anshan Iron & Steel Group Corporation has formed a joint venture company with Tessen of Germany, and Benxi Iron & Steel Group has formed a joint venture company with Posco (Pohang Iron & Steel) of Korea; Chinese companies will be able to adopt new technology through these joint ventures. Although the construction cost of a Nippon Steel Corporation plant in Kitakyushu is estimated at approximately 2 trillion yen, the cost will be one quarter of that if construction takes place in China. This can be interpreted to mean that Nippon Steel Company has been forced to invest in China by the growing "demand capacity" of the Chinese market. Nippon Steel Company will begin to produce material through a joint venture with Shanghai Baosteel. Competition within the Chinese steel industry has intensified since 1997, when the Chinese government abandoned its steel industrial policy and allowed production to increase until it spiralled out of control.

4.3. Firms in Guangzhou

4.3.1. Type I: company T

Company T started its business in China with the production of air-conditioner parts, a business not related to automobile production. In China, local companies have been catching up relatively quickly with the technological level of Japanese companies by introducing CAD (Computer-Aided Design) / CAM (Computer-Aided Manufacturing) into metal-molding production. Company T's products must meet Japanese quality levels and Chinese price levels at the same time. A large number of machines at a price level between 50 billion and 20 billion yen are needed for equipment investment. In 2003, Company T started to produce auto parts as well as electronic equipment parts. As the company does not specialize only in auto parts, the plant did not need to be located close to an automobile production site.

4.3.2. Type I: company HL

Company HL produces key sets for four-wheeled and two-wheeled vehicles. Magnesium and zinc are the key inputs. Plastic die-casting also plays an important role in the production process. HL intends to sell its auto parts products to Nissan and Toyota, even though the firm is currently dependent upon the Honda group in a *keiretsu* relationship. The minimum production level required to attain economies of scale is not clear, since HL produces a variety of auto parts besides key sets.

4.3.3. Type II: company HA

Company HA is one of Honda's in-house departments, which trades genuine Honda auto parts. Because it is a trading company, HA does not possess fixed assets, except for office installations. For this reason, the company can enter Honda's investment markets. HA has no economies of scale.

4.3.4. Type III: company M

Company M produces plastic parts and components, such as engine covers, for Honda. Production. Processes include die-casting, painting, and assembly.

The company was required to increase its production in order to supply Honda's new investment in Wuhan City and to support Honda's new plant operation in Guangzhou. Company M purchases plastic materials from two companies, including Mitsui Chemical, to keep its prices competitive. The company uses large-scale machines, which enable the firm to attain economies of scale at the production level of approximately 400,000 Honda automobiles.

4.3.4. Type III: company FT

Company FT produces clutches, brakes and frames. Production processes include metal stamping, welding, painting, and assembly. Maintenance of molds and a method of three-dimensional measuring that guarantee precision, are key factors for production. The minimum production level of automobiles required to attain economies of scale for this firm is 100,000.

5. Conclusions and summary

This article developed a "flowchart approach to industrial cluster policy" by proposing sufficient conditions for the formation of the type of industrial clusters typical of the manufacturing industry in Asia, and providing theoretical support for the sufficiency of these conditions to enhance regional economic growth. The theory, summarized in Figure A.4, emphasizes the importance of quasi-public goods, increasing returns to scale, and scale economies. The typical formation pattern of industrial clusters in East Asia was modeled by defining "quasi-public goods," and it was shown that industrial cluster policy enhances economic growth under a production function of "increasing returns to scale." Critical amounts of production necessary for "scale economies," which affect firms' decision whether or not to invest in clusters, were also demonstrated.

The concepts of quasi-public goods, increasing returns to scale, and economies of scale are crucial to the theoretical development presented here. Sufficient conditions for cluster development are the establishment of industrial zones, the building of capacity, and the invitation of anchor firms together with their related firms. First, industrial zones and capacity such as physical infrastructure, institutions, and human resources are quasi-public goods, provided by both organizations in the quasi-public sector and firms in the private sector. Second, the quasi-public goods offered by industrial cluster policy, which provides industrial zones and capacity, can enhance regional economic growth in cases where an anchor firm operates under increasing returns to scale. Markets in China are at an early stage of development, but large enough for anchor companies to attain increasing returns to scale. Third, the minimum optimal size of production necessary to attain economies of scale depends on the size of fixed capital of anchor firms' related companies.

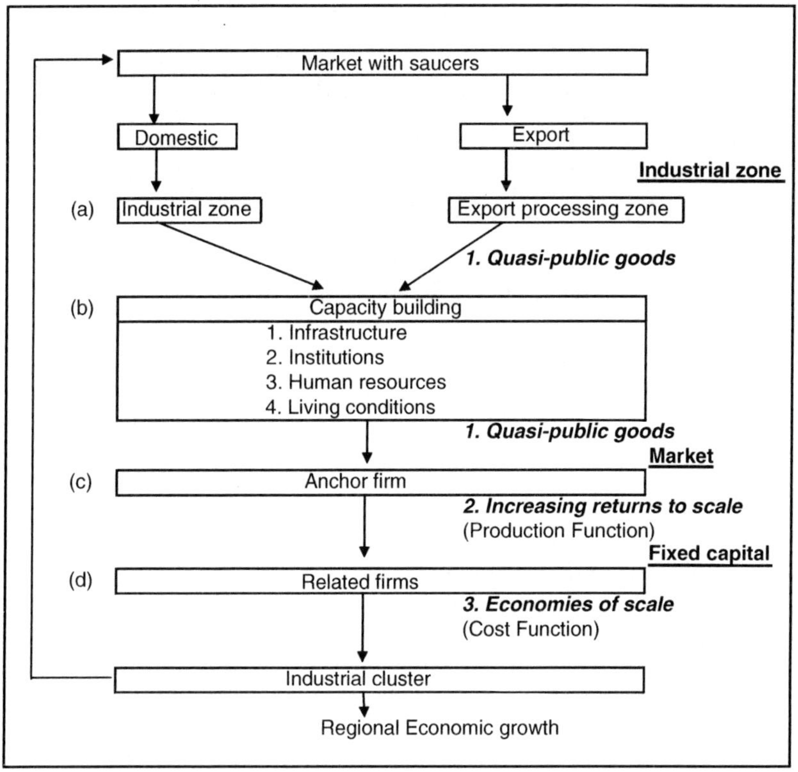

Figure A.4 An industrial cluster formed by an anchor firm
Source: author.

The flowchart approach to industrial cluster policy emphasizes the importance of the ordering and timing of policy measures. The flow of policy implementation is first to establish an industrial zone, then to invite an anchor company, and then to promote the anchor firm's related companies to invest in the industrial zone. In the later stages of clustering, the recipient country's government must reduce its role in order to promote competition. The central government should at that time transfer greater authority to local governments and make more use of the quasi-public sector (public corporations and state enterprises). these measures will result in a quasi-public sector that is likely to supply quasi-public goods (Figure A.2). The improvement and expansion of network formation in Asia by both multinational corporations and the quasi-public sector are thus important steps toward the upgrading of Asia's industrial environment.

Acknowledgements

The author is greatly indebted to M. Fujita, R. Kada, H. Kim, J. W. Longworth, C. Nakajima, Y. Maruyama, K. Suzumura, and M. Tsuji for helpful suggestions and constructive comments on an earlier draft. Responsibility for errors, of course, remains with the author.

Appendix 1: practical considerations

Some economists are critical of a high degree of centralization of governmental power. Friedman (1962) argues that national public goods often serve the benefit of a particular group or result in economic waste, contrary to the intentions of the central government.

Sugioka (1976) advocates "regionalism." This means that governmental power should be decentralized; this is one of the most important topics of discussion in Japan at present.

Little concrete consideration has been given to the question of what kinds of goods and services should be supplied by: (1) local governmental units or (2) entities that belong to neither the public nor private sectors.

The importance of that non-public, non-private sector seems to have been increasing recently. In order to clarify the role of that sector, the concept of quasi-public optional goods is useful. W, Y and T are defined as quasi-public optional goods for the purposes of the above analysis.

The sets, M_b which represent peoples' needs or the structures of their optionality, are used below.

Types of quasi-public good include:

1. Quasi-public goods that are demanded by most local citizens in a city. Sets M_b, consist of such citizen demands (e.g. parks in the cities). From the viewpoint of equity, local government units should supply such goods.
2. Quasi-public goods supplied by organizations in the sector that belong to neither the public nor private sector.

This latter type of good has recently become important, but has not been analyzed sufficiently. Suppose that farmers in a country require many kinds of quasi-public goods. Quasi-public optional goods cannot be neglected. It is not appropriate for the local governmental unit to supply these, because that woud represent unfair treatment. In general, private firms will not give priority to quasi-public goods in terms of profitability. This leaves other institutions to fill the need for this type of good. For example, agricultural cooperatives can be considered as economic units that supply quasi-public goods. Examples include:

T: Establishment and management of water facilities,
T: Wire broadcasting,

T: Research on special products (e.g. flowers),
T: Information about markets, and
Y: Roads for agricultural use.

This sector may supply some kinds of quasi-public optional goods as "local private collective (production or consumption) goods." Of course, further analysis of this sector is required to clarify the role of cooperatives.

Appendix 2

Individual concern versus collective concern:

Millerton (1972) defined this dynamic, and this is similar to Buchanan's (1965) statement: "The interesting cases are those goods and services, the consumption of which involves some 'publicness,' where the optimal sharing group is more than one person or family but smaller than an infinitely large number. The range of 'publicness' is finite."

Appendix 3

Proof: Non-rivalness requires:

$$0 \leq t^i_f \leq 1,$$

then:

$$t^i_f T_f.$$

Define:

$$T^i_f = t^i_f T_f.$$

The constraints on T will be:

$$T^i_f \leq T_f,$$

and these are the same as those of W.

Appendix 4

Proof: When the constraints with respect to W and Y are interpreted, only the differences between w^i_b and y^i_d, which are constant, may be considered. If the possibilities are:

$$\sum_i w^i_b > 1,$$

then the constraints correspond to W. If the following must hold:

$$\sum_i w^i_b \leq 1,$$

then these correspond to Y. Both of these equations are equivalent in the process of calculating optimality conditions. The interpretation of these may only be changed according to w'_b. Similarly, the relationship between X and V can be discussed. Attention is thus concentrated on V and W.

Appendix 5

$$\text{Max } U\,[u^1(V;W^1;Z^1),\ldots,u^s(V;W^s;Z^s)]\,, \tag{1}$$

Subject to:

$$F\,(V,W,Z) \leq 0 \tag{2}$$

$$W^i_b \leq w^i_b\,W_b,\; b = 1,\ldots,k,\; i = 1,\ldots,s, \tag{3}$$

$$\sum_i Z^i_e \leq Z_e,\; e = 1,\ldots,n, \tag{4}$$

$$V,\, W^i,\, W,\, Z^i,\, Z \geq 0\,. \tag{5}$$

Let α, β^i_b, and γ_e represent the multipliers associated with constraints (2), (3) and (4), respectively.

Appendix 6

Proof: The Kuhn-Tucker conditions with respect to W are:

$$A = (\partial U/\partial u^i)(\partial u^i/\partial W^i_b) - \beta^i_b \leq 0,\; A\cdot W^i_b = 0,\; W^i_b \geq 0,$$
$$i = 1,\ldots,s,\quad b = 1,\ldots,T. \tag{1}$$

$$B = \sum_i \beta^i_b - \alpha\,.\,w^i_b\,.\,\partial F/\partial w_b \leq 0,\; B\cdot w_b = 0,\; w_b \geq 0 \tag{2}$$

$$C = -(W^i_b - w^i_b W_b) \geq 0,\; C\cdot\beta^i_b = 0,\; \beta^i_b \geq 0 \tag{3}$$

Consider the following definitions:

$$M_b = \{i \mid W^i_b/w^i_b = \max\,(W^1_b/w^1_b,\ldots,W^s_b/w^s_b)\}$$

$$M_t = \{1,\ldots,s\}$$

$$N_b = M_t - M_b$$

The Sets M_b cannot be empty, since the constraint (3) must be binding for some persons in order to have a rational allocation of resources. Thus, the multipliers β^i_b (for $i \in M_b$) become positive. The strict inequalities,

$$-\,(W^i_b - w^i_b\,W_b) < 0,\text{ for } i \in N_b,$$

must hold in (3). Therefore, from (3),

$$\beta^i_b = 0,\text{ for } i \in N_b\,.$$

Notice that whether the individual belongs to M_b or N_b also depends on W^1_b. In this process, W^1_b plays an important role. Suppose that some set M_b

consists of a single element. That is, consider the extreme case where the maximum is only one person. The strict inequality must hold in (3) except for that person. Then,

$$\partial U / \partial u^i \cdot \partial u^i / \partial W^i_b = \alpha \cdot W^i_b \cdot \partial U / \partial W_b, \; i \in M_b.$$

Appendix 7

Model

Following Romer (1990), the production function of the firms in industrial zones inGuangzhou imitating firms in country 1 may be written as:

$$Y_2 = A_2 L_2^{1-\alpha} \sum_{j=1}^{n1} X_{2j}^{\alpha},$$

where $0<\alpha<1$, n_1 is the number of varieties of intermediate goods, A_2 is a productivity parameter, L_2 is labor input, and X_{2j} is the use of the *j*th type of intermediate good. The government is assumed to run a balanced budget financed by a proportional tax at a rate of *t*. Here, an innovator in country 1 is an intermediate goods producer. An imitator in country 2 is a final goods producer.

The profit of the producer of final goods in country 2 (π_{2F}) is:

$$\pi_{2F} = Y_2 - wL_2 - \sum_{j=1}^{n1} P_j X_{2j},$$

where *w* is the wage rate, and P_j is the price of the intermediate good *j*. The condition to maximize the profit is:

$$X_{2j} = L_2 (A_2 \alpha / P_j)^{1/(1-\alpha)} \qquad (1)$$

Next, we assume monopolistic competition in intermediate goods markets. The profit of intermediate goods producers in country 2 is:

$$\pi_{M2j} = P_j (X_{2j}).\, X_{2j} - X_{2j} \qquad (2)$$

The solution for the monopoly price is:

$$P_j = 1/\alpha \qquad (3)$$

Hence, if (1) and (3) into (2) are used for substitution into P_j and X_{2j}, then:

$$\pi_{M2j} = \{(1-\alpha)/\alpha\} L_2 A_2^{1/(1-\alpha)} \alpha^{2/(1-\alpha)} \qquad (4)$$

It may be assumed that the business is free to imitate the *j*th intermediate good. Thus, anyone can pay the imitation cost *v* to secure (in units of *Y*) the net present value $V_j(T)$ at time T. This can be obtained by time Υ from T to infinity:

$$V_j(T) = \int_T^{\infty} \pi_{M2j}\, e^{-r\,(\Upsilon - T)} d\Upsilon\,, \tag{5}$$

where r is the interest rate and is constant. The free-entry condition requires that:

$$V_j(T) = \nu. \tag{6}$$

Equations (5) and (6) give:

$$\pi_{M2j} / V_j = r,$$

that is:

$$r = (L_2 / \nu)\,\{(1-\alpha) / \alpha\}\, A_2^{1/(1-\alpha)} \alpha^{2/(1-\alpha)} \tag{7}$$

It may be assumed that households maximize utility over an infinite horizon:

$$U = \int_0^{\infty} \{(c^{1-\delta} - 1) / (1-\delta)\}\, e^{-\rho\gamma}\, d\gamma,$$

where ρ is the constant rate of time preference, δ is the coefficient of risk aversion, γ is time, and c is consumption (Here, the rate of population growth is assumed to be 0).

The growth rate of consumption is

$$\dot{c} / c = (1 - \delta)\cdot(r - \rho) \tag{8}$$

Assuming balanced growth, Equations (7) and (8) yield the growth rate of country 2 as follows:

$$g_0 = \dot{c} / c = \dot{Y} / Y,$$

$$= (1 / \delta)\,[\,(L_2 / \nu)\{\,(1-\alpha) / \alpha\}\, A_2^{1/(1-\alpha)} \alpha^{2/(1-\alpha)} - \rho] \tag{9}$$

(2) The Case of Multinational Companies Investing in Export-Processing Zones

In Case 2, firms from country 1 decide to invest in an export-processing zone (EPZ) in country 2.

Let G represent the total of government investment in an export-processing zone as explained in Barro and Sarai-i-Martin (1995). The production function of a final good Y of the firms from country 1 in the EPZ is:

$$Y_1 = A_2 L_1^{\gamma} (\sum_{j=1}^{n1} X_{1j}^{\alpha}) G^{\beta},$$

where L_1 is labor input, and X_{1j} is the use of the j th type of intermediate good. The before-tax profit for the firms in the EPZ of the final good is:

$$\pi_{2F} = \theta Y_1 - w L_1 - \sum_{j=1}^{n1} P_{2j} X_{1j},$$

where w is the wage rate, and θ is country risk in the EPZ for firms of country 1. This parameter was explained in the previous section and is different from the productivity parameter A_2.

Whether multinationals make a decision to invest in the EPZ depends on whether or not the country risk θ is lower than the threshold rate. The value of this threshold is determined by the key variable of tax rate t, and this depends on the government of country 2. The lower the tax rate set by the government of country 2, the more easily the firms of country 1 can invest in the EPZ of country 2. The importance of the role of government in providing quasi-public goods is emphasized below.

Here, the case of foreign investment of firms from country 1 in the EPZ is considered. In terms of the currency of country 2 before taxation, the profit of firms from country 1 relative to an intermediate good is:

$$\pi_{M1j} = \{(1-\alpha)/\alpha\}\,(\theta A_2)^{1/(1-\alpha)}\,\alpha^{2(1-\alpha)}L_1.$$

Then, the tax revenue for the government of country 2 is:

$$t\pi_{M1j} \tag{10}$$

The total flow of monopoly profit to country 2 from the imitation of a new product of country 1, and the invitation to the EPZ in country 2, is the sum of the amounts given by Equations (4), (9) and (10):

$$R_{2j} = \{(1-\alpha)/\alpha\}\,A_2^{1/(1-\alpha)}\,\alpha^{2(1-\alpha)}L_2 + t\{(1-\alpha)/\alpha\}\,X_{1j} + \frac{\gamma}{\alpha^2}X_{1j},$$

where the second term is profit tax revenue, and the third term is labor income per industry.

It is assumed that the government of country 2 has spent cost λ to invite firms from country 1 to the EPZ. So, the total cost of country 2 is:

$$\nu + \lambda.$$

The rate of return to country 2 is:

$$R_{2j}/(\nu + \lambda).$$

Then, the growth rate of country 2 is:

$$g_1 = (1/\delta)\,\{R_{2j}/(\nu + \lambda) - \rho\} \tag{11}$$

Setting $g_1 - g_o = K - 1$, where:

$$K = \frac{\nu}{\lambda}\cdot\alpha^{\frac{2\alpha\beta}{(1-\alpha-\beta)(1-\alpha)}}(\theta A_2)^{\frac{\beta}{(1-\alpha-\beta)(1-\alpha)}}\{n(1-\alpha-\gamma)\}^{\frac{\beta}{1-\alpha-\beta}}\frac{L_1^{\frac{\gamma}{1-\alpha-\beta}}}{L_2}\times\left\{t+\frac{\gamma}{\alpha(1-\alpha)}\right\}t^{\frac{\beta}{1-\alpha-\beta}}-1$$

it can be seen that the sign of $g_1 - g_o$ depends of the sign of K. The larger $\frac{\nu}{\lambda}$ and n are, the more probable it is that the sign of $g_1 - g_o$ will be positive.

In other words, governments should invite FDI to EPZs if the cost of invitation is much cheaper than that of imitation, or if the number of intermediate goods is large.

We assume that a government runs a balanced budget financed by a proportional tax at rate t. That is:

$$t\pi_{2F} = G.$$

The conditions for the maximization of the firm in country 2 in the EPZs are:

$$wL_1 / n_1 = \gamma(\theta A_2 L_1{}^{\gamma} X_{1j}{}^{\alpha} G^{\beta}), \tag{12}$$

and:

$$P_j = \alpha\theta A_2 L_1{}^{\gamma} X_{1j}{}^{\alpha} G^{\beta} \tag{13}$$

Using equation (3), (12), and (13), the following wage revenue of the jth industry is obtained:

$$wL_1 / n = \gamma X_{1j} / \alpha^2 \tag{14}$$

We can derive the conditions under which a recipient country should set the tax rate lower in order to enhance the aggregate growth rate by inviting multinational companies and reducing the profit tax. The partial derivative of K with respect to t is:

$$\frac{\partial K}{\partial t} = H \cdot t^{a-1} \cdot \left\{ t + \frac{\beta\gamma}{\alpha(1-\alpha)^2} \right\} \cdot \frac{1-\alpha}{1-\alpha-\beta},$$

where:

$$H = \frac{L_1{}^{\frac{\gamma}{1-\alpha-\beta}}}{L_2} \frac{\nu}{\lambda} \alpha^{\frac{2\alpha\beta}{(1-\alpha-\beta)(1-\alpha)}} (\theta A_2)^{\frac{\beta}{(1-\alpha-\beta)(1-\alpha)}} \{n(1-\alpha-\gamma)\}^{\frac{\beta}{1-\alpha-\beta}}$$

The sign of $\frac{\partial K}{\partial t}$ is negative in the case of $1-\alpha-\beta < 0$ and $\alpha < 1$(the production function of this case has increasing returns to scale).

Suppose that the production function of a final good Y of the firms from country 1 in the EPZ has increasing returns to scale; that is, $1 < \alpha + \beta$. Then, national income is increased by the reduction of the profit tax. The lower the tax rate, the higher the growth rate.

References

Barro, R., and Xavier Sala-i-Martin. *Economic Growth.* New York: McGraw-Hill, 1995.

Buchanan, J. M. "An Economic Theory of Clubs." *Economica,* vol.32, 1965.

Davis, O.A., and A. B. Winston. "On the Distinction Between public and Private Goods." *American Economic Review* vol.57, 1967.

Dorfman, R. "General Equilibrium with Public Goods." In *Public Economics,* ed. J. Margolis and H. Guitton. Macmillan, 1969.

Frank, Jr., C. K. *Production Theory and Indivisible Commodities*. Princeton, New Jersey: Princeton University Press, 1969.

Friedman, M. *Capitalism and Freedom*. Chicago: University of Chicago Press, 1962.

Gomory, R. E., and W. J. Baumol. "Integer Programming and Pricing." *Econometrica* vol.28, 1960.

Grossman, G. E., and E. Helpman. *Innovation and Growth in the Global Economy*. Cambridge, MA: MIT Press, 1991.

Hamada, K. "An Economic Analysis of the Duty-Free Zone." *Journal of International Economics* no.4, 1974: 225–241.

Hayami, Y., and S. Yamada. "Agricultural Research Organization in Economic Development." In *Agriculture in Development Theory*, ed. L. G. Reynolds. New Haven: Yale University Press, 1975.

Holtermann, S. E. "Externalities and Public Goods." *Economica*, vol.39, 1972.

James, E. "Joint Products, Collective Goods, and External Effects: Comment." *Journal of Political Economy*, vol. 77, 1969.

Markusen, A. "Sticky Places in Slippery Space: A Typology of Industrial Districts." *Economic Geography*, vol. 72, 1996: 293–313.

Milleron, J. C. "Theory of Value with Public Goods: A Survey Article." *Journal of Economic Theory*, vol. 4, 1972.

Minasian, J. R. "Television Pricing and the Theory of public Goods." *Journal of Law and Economics*, vol. 7, 1964.

Mitsui, I. "Industrial Cluster Policy and Regional Development in the Age of Globalization." *30th International Small Business Congress in Singapore* 2003.

Musgrave, R. *Fiscal System*. New Haven, Conn.: Yale University Press, 1969.

Musgrave, R. "Provision for Social Goods," In *Public Economics*, ed. J. Margolis and H. Guitton. Macmillan, 1969.

Ng, Yew-Kwang. "The Economic Theory of Clubs: Pareto Optimality Conditions." *Economica*, vol. 40, 1973.

Porter, M. E. *The Competitive Advantage of Nations*. New York: Free Press, 1990.

Romer, P. M. "Endogenous Technological Change." *Journal of Political Economy*, vol. 98, no. 5, 1990: S71–S102.

Samuelson, P. A. "The Pure theory of Public Expenditure." *Review of Economics and Statistics*, vol. 36, 1954: 387–389.

Samuelson, P. A. "Diagrammatic Exposition of Theory of a Public Expenditure." *Review of Economics and Statistics*, vol. 37, 1955: 350–356.

Index

Introductory Note

References such as "178–9" indicate (not necessarily continuous) discussion of a topic across a range of pages, whilst "123f5.1" indicates a reference to figure 5.1 on page 123, "134t5.1" a reference to table 5.1 on page 134 and "145n2" a reference to endnote 2 on page 145 (with "nn" used where multiple notes are referenced). Wherever possible in the case of topics with many references, these have either been divided into sub-topics or the most significant discussions of the topic are indicated by page numbers in bold.